Functional Foods II
Claims and Evidence

Functional Foods II
Claims and Evidence

Edited by

Judy Buttriss
The British Nutrition Foundation, London, UK

Michael Saltmarsh
Inglehurst Foods, Alton, Hampshire, UK

ROYAL SOCIETY OF CHEMISTRY

The Proceedings of a joint conference held by the British Nutrition Foundation and the Food Chemistry Group of The Royal Society of Chemistry on 14–15 April 1999 at Wye College, University of London, Kent, UK

Special Publication No. 248

ISBN 0-85404-789-1

A catalogue record of this book is available from the British Library

Published by The Royal Society of Chemistry,
Thomas Graham House, Science Park, Milton Road, Cambridge CB4 0WF, UK

For further information see our web site at www.rsc.org

Typeset by Computape (Pickering) Ltd, Pickering, North Yorkshire, UK
Printed and bound in Great Britain by MPG Books Ltd, Bodmin, Cornwall

Editors' Introduction

These are the proceedings of a joint conference held by the British Nutrition Foundation and the Food Chemistry Group of the Royal Society of Chemistry on 14th–15th April 1999 at Wye College, University of London, in Kent.

This is the second conference organised by this group on the subject of functional foods. Since the first conference in 1997, the food industry has continued to expand its development of foods with positive nutritional benefits so that the pre- and probiotic foods which dominated the market in the early days are now joined by foods high in flavonoids, phytoestrogens and a range of other phytochemicals. It was thus timely to provide an opportunity to review the data that supports these products.

The aims of the conference were to bring together those involved in academic and industrial research and product development to review progress in two areas pertinent to the growth of the market in functional foods. These were:

- evidence for the benefit of physiologically functional ingredients for human health
- the consumer and regulatory background against which the market is developing.

The programme included 7 plenary lectures and 19 contributed papers. It was not possible to reproduce one of the plenary lectures in this volume. Mr Roger Manley, who gave the first paper on the regulatory background in the UK had to go into hospital after the conference and under the circumstances it was unreasonable to expect him to provide a copy of his paper. Also, following the conference, family problems prevented Professor Setchell from providing a paper based on his lecture. The editors would like to record their appreciation to Mr Manley and Professor Setchell for their contributions to the conference and to all the lecturers for their participation in enthusiastic debate both in and outside the conference hall. It is the continuation of this debate which will inform the discussion of the real value of functional ingredients in the coming years.

We are grateful for financial support which enables conferences like these to be run at fees which allow the participation of academic researchers from both the UK and overseas. Support for the social programme for this conference was contributed by J Sainsbury plc, Meat and Livestock Commission and Yakult UK. Other support was received from Eurofins Scientific, Leatherhead Food RA, Lines of Communication, National Dairy Council, Nestlé LC1, Roche Products Ltd, SmithKline Beecham and Van den Burgh Foods. Contributions to the administrative support were received from British Sugar plc, Cadbury Limited, Nestlé UK Limited, St Ivel Limited, Unilever plc and Weetabix Limited.

Finally we would like to thank Priscilla Appiah of the British Nutrition Foundation for ensuring that the conference ran smoothly and for her valuable help with the production of the conference proceedings.

We hope that publication of these proceedings will enable a wider audience to share the scientific content of the meeting.

Judy Buttriss and
Mike Saltmarsh

Contents

Contributors

P J Aggett
University of Central Lancashire, Lancashire Postgraduate School of Medicine and Health, Preston, Lancashire, UK

A Alhamdan
Human Nutrition, University of Southampton, Bassett Crescent East, Southampton, SO16 7PX, UK

H S Aro
Agricultural Research Centre of Finland, Food Research, Jokioinen, Finland

M Ashwell
Ashwell Associates, Ashwell, Hertfordshire, UK

A-M Aura
VTT Biotechnology & Food Research, PO Box 1500, 02044 VTT, Finland

K Barnes
CSL Food Science Laboratory, Norwich Research Park, Colney, Norwich, NR4 7UQ, UK

S Blum
Nestec Ltd., Nestlé Research Centre, Vers-chez-les-Blanc, PO Box 44, 1000 Lausanne 26, Switzerland

F Bornet
Eridania Béghin-Say, Vilvoorde Research & Development Centre, Nutrition & Health Service, Vilvoorde, Belgium

D Brassart
Nestec Ltd., Nestlé Research Centre, Vers-chez-les-Blanc, PO Box 44, 1000 Lausanne 26, Switzerland

E J Brink
TNO Nutrition & Food Research Institute, PO Box 360, 3700 AJ Zeist, The Netherlands

J M M Chin a Paw
Division of Human Nutrition & Epidemiology, Wageningen Agricultural University, Dreijenlaan 1, 6703 HA Wageningen, The Netherlands

J K Collins
Department of Microbiology and Medicine, University College Cork, Cork, Ireland

I Corthésy-Theulaz
Nestec Ltd., Nestlé Research Centre, Vers-chez-les-Blanc, PO Box 44, 1000 Lausanne 26, Switzerland

M L Cross
Milk and Health Research Centre, Institute of Food, Nutrition and Human Health, Massey University, Private Bag 11222, Palmerston North, New Zealand

A Damant
CSL Food Science Laboratory, Norwich Research Park, Colney, Norwich, NR4 7UQ, UK

C P G M de Groot
Division of Human Nutrition & Epidemiology, Wageningen Agricultural University, Dreijenlaan 1, 6703 HA Wageningen, The Netherlands

N de Jong
Division of Human Nutrition & Epidemiology, Wageningen Agricultural University, Dreijenlaan 1, 6703 HA Wageningen, The Netherlands

A T Diplock
International Antioxidant Research Centre, GKT Guy's Hospital, London, UK

A Donnet-Hughes
Nestec Ltd., Nestlé Research Centre, Vers-chez-les-Blanc, PO Box 44, 1000 Lausanne 26, Switzerland

A G Dulloo
Institute of Physiology, University of Fribourg, Rue du Musée 5, CH-1700 Fribourg, Switzerland

C Dunne
Department of Microbiology, National Food Biotechnology Centre, University College Cork, Cork, Ireland

M Feeney
Department of Microbiology, University College Cork, Cork, Ireland

E Fern
Nestec Ltd., Vevey, Switzerland

M Fewtrell
MRC Childhood Nutrition Research Centre, Institute of Child Health, London, WC1N 1EH, UK

G Fitzgerald
Department of Microbiology, University College Cork, Ireland

P Forssell
VTT Biotechnology & Food Research, PO Box 1500, 02044 VTT, Finland

G R Gibson
Department of Food Science and Technology, University of Reading, UK

H S Gill
Milk and Health Research Centre, Institute of Food, Nutrition and Human Health, Massey University, Private Bag 11222, Palmerston North, New Zealand

M Gordon
University of Reading, Dept. of Food Science, PO Box 226, Whiteknights, Reading RG6 6AP, UK

A B Hanley
CSL Food Science Laboratory, Norwich Research Park, Colney, Norwich, NR4 7UQ, UK

H F J Hendriks
TNO Nutrition and Food Research Institute, PO Box 360, 3700 AJ Zeist, The Netherlands

G J Hiddink
Dutch Dairy Foundation on Nutrition and Health, PO Box 6017, 3600 HA Maarssen, The Netherlands

A Hietanen
Agricultural Research Centre of Finland, Food Research, 31600 Jokioinen,
Finland

J Hiidenhovi
Agricultural Research Centre of Finland, Food Research, 31600 Jokioinen,
Finland

R Huopalahti
University of Turku, Department of Biochemistry & Food Chemistry, 20014
Turku, Finland

A. Husband
Novogen Ltd, 140 Wicks Road, North Ryde, NSW, 2113, Australia

N Jotcham
Yakult UK Ltd., 12-16 Telford Way, Westway Estate, Acton, London, W3,
UK

M Kärkkäinen
Department of Applied Chemistry & Microbiology, Division of Nutrition, PO
Box 27, 00014, University of Helsinki, Finland

S Karppinen
VTT Biotechnology & Food Research, PO Box 1500, 02044 VTT, Finland

B Kiely
National Food Biotechnology Centre, University College Cork, Cork, Ireland

T Kiiskinen
Agricultural Research Centre of Finland, Animal Breeding, Jokioinen,
Finland

M Kilpi
Agricultural Research Centre of Finland, Food Research, 31600 Jokioinen,
Finland

H Korhonen
Agricultural Research Centre of Finland, Food Research, 31600 Jokioinen,
Finland

R Korpela
Valio Ltd R&D Centre (Foundation for Nutrition Research), PO Box 30,
00039 Helsinki, Finland

B Kremers
Mars Inc., Veghel, The Netherlands

C Lamberg-Allardt
Department of Applied Chemistry & Microbiology, Division of Nutrition, PO
Box 27, 00014, University of Helsinki, Finland

H Link-Amster
Nestec Ltd., Nestlé Research Centre, Vers-chez-les-Blanc, PO Box 44, 1000
Lausanne 26, Switzerland

K Liukkonen
VTT Biotechnology & Food Research, PO Box 1500, 02044 VTT, Finland

G T MacFarlane
Medical School, Nine Wells Hospital, University of Dundee, UK

D P Makris
Department of Biological Sciences, Wye College, University of London, Wye,
Ashford, Kent, TN25 5AH, UK

R Manley OBE
Cheshire Trading Standards, Cheshire County Council, County Hall, Chester,
Cheshire, CH1 1SF, UK

S Mantere-Alhonen
University of Helsinki, Department of Food Technology, 00014 Helsinki,
Finland

S McBrearty
Teagasc, Dairy Products Research Centre, Morrepark, Fermoy, Co. Cork,
Ireland

J McBride
CSL Food Science Laboratory, Norwich Research Park, Colney, Norwich,
NR4 7UQ, UK

A L McCartney
Department of Food Science and Technology, University of Reading, UK

S McEvoy
St James' Hospital – Dublin, Nutrition Laboratory, Dept. of Clinical Medicine, Trinity Centre for Health Sciences, Dublin 8, Ireland

G W Meijer
Unilever Nutrition Center, Unilever Research, Vlaardingen, The Netherlands

L Murphy
Department of Microbiology, University College Cork, Cork, Ireland

J R Neeser
Nestec Ltd., Nestlé Research Centre, Vers-chez-les-Blanc, PO Box 44, 1000
Lausanne 26, Switzerland

S L Oehlschlager
CSL Food Science Laboratory, Norwich Research Park, Colney, Norwich,
NR4 7UQ, UK

S O'Halloran
Department of Microbiology, University College Cork, Cork, Ireland

L O'Mahony
Department of Microbiology, National Food Biotechnology Centre, University College Cork, Cork, Ireland

G O'Sullivan
Department of Surgery, Mercy Hospital, Cork, Ireland

S Panula
Agricultural Research Centre of Finland, Animal Breeding, Jokioinen,
Finland

A Pfeifer
Nestec Ltd., Nestlé Research Centre, Vers-chez-les-Blanc, PO Box 44, 1000
Lausanne 26, Switzerland

J-M Pihlava
Agricultural Research Centre of Finland, Food Research, 31600, Jokioinen,
Finland

B Pöpping
CSL Food Science Laboratory, Norwich Research Park, Colney, Norwich,
NR4 7UQ, UK

K Poutanen
VTT Biotechnology & Food Research, PO Box 1500, 02044 VTT, Finland

T Purtsi
Agricultural Research Centre of Finland, Food Research, 31600, Jokioinen,
Finland

M M Raats
(Formerly Research Manager, Health Education Authority), Unit of Health

Care Epidemiology, University of Oxford, Institute of Health Sciences, Old Road, Oxford OX3 7LF, UK

B A Rabiu
Department of Food Science and Technology, University of Reading, UK

M Rinta-Koski
University of Helsinki, Department of Food Technology, 00014 Helsinki, Finland

M Roberfroid
Université Catholique de Louvain, Ecole de Pharmacie, Brussels, Belgium

R P Ross
Teagasc, Dairy Products Research Centre, Morrepark, Fermoy, Co. Cork, Ireland

J T Rossiter
Department of Biological Sciences, Wye College, University of London, Wye, Ashford, Kent, TN25 5AH, UK

K J Rutherfurd
Milk and Health Research Centre, Institute of Food, Nutrition and Human Health, Massey University, Private Bag 11222, Palmerston North, New Zealand

E-L Ryhänen
Agricultural Research Centre of Finland, Food Research, Jokioinen, Finland

E J Schiffrin
Nestec Ltd., Nestlé Research Centre, Vers-chez-les-Blanc, PO Box 44, 1000 Lausanne 26, Switzerland

J Scott
Trinity College, Dublin, Ireland

A L Servin
CJF 94.07 INSERM, UFR de Pharmacie, Université Paris XI, 92296 Chatenay-Malabry, France

K Setchell
Clinical Mass Spectrometry, Dept. of Paediatrics, 3333 Burnet Avenue, Cincinnati, Ohio 45229, USA

F Shanahan
Department of Microbiology and Medicine, University College Cork, Cork, Ireland

C Shortt
Yakult UK, Acton, London W3 7XS, UK

P J Simpson
Teagasc, Dairy Products Research Centre, Morrepark, Fermoy, Co. Cork, Ireland

R Smith
CSL Food Science Laboratory, Norwich Research Park, Colney, Norwich, NR4 7UQ, UK

S Southon
Insititue of Food Research, Norwich Research Park, Colney, Norwich NR4 7UA, UK

C Stanton
Teagasc, Dairy Products Research Centre, Morrepark, Fermoy, Co. Cork, Ireland

M Taipale
Agricultural Research Centre of Finland, Food Research, 31600, Jokioinen, Finland

U Teuri
Valio Ltd R&D Centre, PO Box 30, 00039, Helsinki, Finland

L Thorpe
Health Education Authority, Trevelyan House, 30 Great Peter Street, London SW1P 2HW, UK

T Tupasela
Agricultural Research Centre of Finland, Food Research, Jokioinen, Finland

W A van Staveren
Division of Human Nutrition & Epidemiology, Wageningen Agricultural University, Dreijenlaan 1, 6703 HA Wageningen, The Netherlands

T van Vliet
TNO Nutrition and Food Research Institute, PO Box 360, 3700 AJ Zeist, The Netherlands

B Viander
Agricultural Research Centre of Finland, Food Research, 31600, Jokioinen, Finland

A F Walker
The Hugh Sinclair Unit of Human Nutrition, Department of Food Science & Technology, University of Reading, RG6 6AP, UK

J Wallace
School of Biomedical Sciences, University of Ulster, Coleraine, Northern Ireland, BT52 1SA, UK

J A Westrate
Unilever Nutrition Center, Unilever Research, Vlaardingen, The Netherlands

J Young
Leatherhead Food RA, Randalls Road, Leatherhead, Surrey, KT22 7RY, UK

1 Setting the Scene

Factors to Consider when Undertaking Clinical Trials for Functional Foods

S. Southon

INSTITUTE OF FOOD RESEARCH, NORWICH RESEARCH PARK, COLNEY, NORWICH, NR4 7UA

1 Introduction

Functional products and ingredients are defined as having disease preventing and/or health promoting benefits in addition to their nutritive value. This definition fits most convincingly with respect to food ingredients with little or no nutritive value at all, but which may elicit beneficial physiological responses, for example, dietary fibres (including oligosaccharides) and live micro-organisms which are purported to maintain a 'healthy' gut micro-flora. Beyond products containing such ingredients, however, I, and many other consumers start to get confused. The definition implies that 'nutritive value' is considered non-functional although many 'functional foods' appear to be products fortified with ingredients which could be obtained from a primary 'non-functional' food source which has only its nutritive value to recommend it. Although industry is being exhorted to produce innovative ingredients rather than producing just another fortified product, the fortified product is still being developed and marketed in the functional food category. Taking calcium as an example, calcium was claimed as the leading 'functional' food ingredient in Japan (the functional food capital of the world) between 1988 and 1995. Perhaps the confusion arises from the fact that the functional food concept is, at present, more of a marketing than a healthy eating/nutritional concept.

I apparently am not the only one to be confused. In several business reviews, I see that companies are being encouraged to take advantage of the new competitive space for functional foods. However, it is stated, since functional foods refuse to sit neatly into existing market segments, or product categories, there is consumer and media confusion: 'the future success of functional foods will rest on the effectiveness of communication strategies throughout the food chain, but especially at the retail and consumer end'.[1] But what is to be communicated? Obviously, the functional food concept needs clarification both within and outwith the industry but the concern with clarifying the concept should not detract from the more important task of clarifying the purported function of the product and the evidence of that functionality.

Although evidence of benefit is not necessary for effective marketing of a product, marketing would be more ethical if benefit could be demonstrated. The approaches for doing this are no different to those associated with elucidating causative associations for any ingredient, any food, any groups of foods or whole diets that are claimed to promote good health.

2 Scientific Approaches

There are five traditional routes for determining potential and actual benefit: epidemiological associations; animal studies; *in vitro* approaches, such as cell and tissue response in culture; small exploratory human studies; and clinical trials. Some researchers assert that clinical trials represent the only gold standard by which to test hypotheses regarding dietary factors and health. In this presentation I shall evaluate the role of clinical trials in addressing questions of efficacy; in particular, issues relating to subjects, substances, and outcomes that can be tested. The following provides a summary of the areas which will be covered in more detail, with specific examples.

3 Subjects

The aim of a functional food or functional ingredient is to promote health and well-being. However, good health is inherently difficult to measure and when examining the preventive properties of specific food components in clinical trials, it is not possible to start with a healthy population with no risk factors and follow them until enough have developed the disease of interest. The sample size required for such a study would be too large and following progress would not be practically or financially feasible. Thus, researchers are forced to study persons who have a full blown disease or who are at high risk of disease. So the 'preventive' angle actually comes in the form of speculation based on the ability of a food or food ingredient to cure or modify disease processes in subjects who have, or are highly likely to have, a disease condition. Most, if not all, studies conducted so far are of this type.

What do we gain from clinical trials using high risk volunteers? If such studies show no response, or an adverse response, we will not know if the treatment might have been beneficial in the majority of individuals without that particular pre-existing disease. On the other hand, if a response is observed, we will not know if non-high risk subjects would derive any benefit.

The clinical trial has been developed to test the efficacy of drugs where a cure or some significant delay in the progression of a disease is being sought. In drug trials, the study starts with a clearly defined patient group with an existing disease state, a clearly defined outcome and a specific drug to which access is under the control of a physician and to which the subject has probably not been exposed before. Patients are randomly assigned to treatments and the answer is clear. Treatment X is better than treatment Y (or is not) in curing, or modifying, a specific disease, when taken over a specific time period and at a specific dose. The time required for effect, the dose at which an effect is seen (or not) and all contra-indications are taken into account.

Now let us consider curative properties in relation to food components or ingredients. With respect to the 'cure' question, the patient (or subject) group can be clearly defined as in the drug trial but, unless the component/ingredient is something that the patient does not consume, dosage will be variable (maybe highly variable), since the researcher is topping-up the habitual diet. In addition, the tissue status of individuals may vary substantially and so response may vary for the same treatment dose, depending on the baseline 'status' of the volunteer, and this makes interpretation of results difficult. The best use of the clinical trial for foods and food ingredients is (as with the drug area) where the ingredient being tested does not form part of the habitual diet and where there is a clear and obvious outcome within the relatively short time span of the trial. The worst use of the clinical trial is where the component is substantially (and variably) present in the habitual diet, where the disease progression is slow to develop and slow to respond (like cancer and cardio-vascular disease) or where biomarkers of response have not been validated, and where results from sick people are intended to be translated into recommendations for all people.

4 Substances

There are also problems of interpretation relating to the substance under investigation. I use the term substance since trials rarely undertake to examine the food product itself. Most published trials have involved a single component, occasionally two, at a single dose. Usually this dose is well above the amounts indicated as beneficial by epidemiological studies of diet-health relationships, or that which might be consumed on an habitual basis by the average consumer. Sometimes the component tested is chemically, (or biologically), synthesised and may differ in nature from the component present in those foods indicated as beneficial in the diet. Results of trials may indicate benefit, no effect or an adverse effect (over a relatively short intervention period) but how is this to be interpreted in the context of the component contained within a particular food, being consumed as part of a complex diet over many years? What if prevention of initiation requires exposure to the preventive agent from childhood onwards? What if there is an early critical period in which the component must be supplied in ample amounts? It is also well established that nutrients and other food components interact substantially with one another. This is not accounted for in the single component (usual), or even single food (rare) trial.

5 Status

As if the problems associated with subjects and substances were not enough, there is also the knotty problem of assessing outcome in terms of health status. This is less of an issue when the question 'Can we cure a disease?' is addressed. However, when we want to know if a disease can be prevented (and health status promoted) by the consumption of specific products, the usefulness of currently employed biomarkers of early initiating events must be taken into account. Measurements are usually applied to tissues readily accessible from human volunteers (*i.e.* the potential target tissues may not be sampled) and measurements of response often employ biomarkers, which may not represent well established events in the initiation or progression of disease.

6 Suggestion

So where do we go from here? Do we abandon the clinical trial? Do we have to rely on epidemiology, studies in laboratory animals, and/or studies *in vitro* as our source of inspiration and information?

An epidemiologically observed association is relevant to the species but is not causative. Nevertheless, I would say that strong and consistent epidemiological associations are good enough evidence for promoting the sale and consumption of those diets and food groups that show the strongest association. Note, I do not refer here to components of diets and foods but to the actual foods. They also provide a pointer for experimentation but we should always consider the possibility that they might lead us up the garden path, particularly with respect to the bioactivity of individual compounds.

Laboratory data from animal and *in vitro* cell studies can be rigorous but may be irrelevant to humans.

Finally, there are *in vitro* and *ex vivo* studies using human cells and tissues. These too can be rigorous but again may be irrelevant to humans or to conditions *in vivo*. They may also be highly misleading, since such studies may report functional responses to compounds that are not absorbed into the body and/or bear no relation to the compound which might be presented to the cells after digestion and absorption in the intact human. The 'phytoprotectants' area of research is an example of a particular minefield with respect to the type of compound used in studies *in vitro*. There are also problems associated with creating physiological, or near physiological, conditions *in vitro* with respect to compound delivery systems, oxygen tension and media composition, and selection of cell type can be crucial to response.

For all their faults, clinical trials can provide additional data, alongside other approaches, to add to our body of knowledge about potential health benefits of foods and food components, but such trials should not be considered the gold standard and should not be over-interpreted. Comments based on results must be limited to the conditions of the trial, taking into account all the potential confounders, and conclusions must be based only upon the substance/ingredient tested, without excessive speculation as to long-term response to consumption of foods containing this substance/ingredient. In the words of one eminent investigator in this field 'For many questions about the role of nutritional factors in primary prevention of diseases of long latency, the only answer lies in intelligent synthesis'.[2]

References

1   M. Heasman, Creating competitive space in global markets. FT Food Business Sector Report, Issue 8, September 1997.
2   G. Block, *Am. J. Clin. Nutr.*, 1995, **62** (suppl): 1517S.

Scientific Concepts of Functional Foods in Europe: Consensus Document

A.T. Diplock[1], P.J. Aggett[2], M. Ashwell[3], F. Bornet[4], E.B. Fern[5], M. B. Roberfroid[6]

[1] INTERNATIONAL ANTIOXIDANT RESEARCH CENTRE, GKT GUY'S HOSPITAL, LONDON, UNITED KINGDOM
[2] UNIVERSITY OF CENTRAL LANCASHIRE, LANCASHIRE POSTGRADUATE SCHOOL OF MEDICINE AND HEALTH, PRESTON, LANCASHIRE, UNITED KINGDOM
[3] ASHWELL ASSOCIATES, ASHWELL, HERTFORDSHIRE, UNITED KINGDOM
[4] ERIDANIA BÉGHIN-SAY, VILVOORDE RESEARCH & DEVELOPMENT CENTRE, NUTRITION & HEALTH SERVICE, VILVOORDE, BELGIUM
[5] NESTEC LTD., VEVEY, SWITZERLAND
[6] UNIVERSITÉ CATHOLIQUE DE LOUVAIN, ECOLE DE PHARMACIE, BRUSSELS, BELGIUM

1 Preface: ILSI Europe's Role

The European Commission Concerted Action on Functional Food Science in Europe (FUFOSE), which is co-ordinated by The International Life Sciences Institute – ILSI Europe, aims to establish a science-based approach for concepts in functional food science. The goal of this Concerted Action has been to set up a multidisciplinary European network:

1. to assess critically the science base required to provide evidence that specific nutrients and food components positively affect target functions in the body;
2. to examine the available science from a function-driven perspective rather than a product-driven one; and

Reproduced from the British Journal of Nutrition, Volume 80, supplement 1, August 1998 with permission of the Nutrition Society and CABI Publishing, Wallingford, Oxon, UK. © Nutrition Society 1999.

3. to reach consensus on targeted modifications of food and food constituents, and options for their application.

This approach led to identifying key partners from Europe's food and agricultural industry, governmental and inter-governmental bodies and the scientific community. This project provided them with an opportunity to exchange ideas and to interact on a neutral platform.

1.1 Theme Groups for Physiological Functions

The First Plenary Meeting on Functional Food Science in Europe: State of the Art was held in Nice, France, in April 1996. Based on the results of this meeting, six major areas in human physiology were selected and corresponding Individual Theme Groups (ITG) were set up and charged with producing theme papers to review critically the science base of the concepts in their respective areas:

- Growth, development and differentiation: a functional food science approach.
- Functional food science and substrate metabolism.
- Functional food science and defence against reactive oxidative species.
- Functional food science and the cardiovascular system.
- Functional food science and gastrointestinal physiology and function.
- Functional food science and behaviour and psychological function.

Each ITG reviewed the published literature to: define the state of the art with respect to specific body systems; assess critically methodologies to characterize and quantify specific related functions; identify and review nutritional options modulating these functions; evaluate potential safety implications related to these nutritional options; identify the role of food technology on nutritional and safety aspects; assess critically the science-base required for providing evidence that specific nutrients positively affect functions; identify areas where further research is required.

The resulting documents were scrutinized in a Second Plenary Meeting held in July 1997 in Helsinki, Finland, and revised by the ITG chairs to include the comments made during that consultation. The final reports of the six ITGs were published in the *British Journal of Nutrition*.[1]

1.2 Theme Group for Technology

An Expert Group on Food Technology was also established to examine the impact and feasibility of food technology on functional food development. Bearing in mind that a function-driven approach rather than a product-driven approach should be adopted, the group considered examples of technologies that are applied to raw materials to improve their quality, the identification of new materials and new processes, and the interactions between processing and functionality.

The group reviewed the impact of food processing on themes such as technological processes to optimize antioxidants, minerals, micro-organisms, carbohydrates and peptides.

The Technology Theme Group was asked to address topics such as the state of the art (impact of processing); source of materials; processing options modulating functionality, including post-harvest technology, unit operations and storage and distribution/packaging; safety implications of materials and processes; process monitoring for functions; and further development and research needs.

These five technology theme papers have been published in Trends in Food Science and Technology.[2]

1.3 Concept of the Consensus Document

The six ITG reviews and that of the Technology Group provided the foundation for this Consensus Document on Scientific Concepts of Functional Foods in Europe. The outline for the document was reviewed by participants at the Second Plenary Meeting in Helsinki, Finland.

The draft document was reviewed by the Steering Committee set up to monitor this project, as well as the ILSI Europe Functional Food Task Force, and by the participants in the Third Plenary Meeting, which was held in Madrid in October 1998. Their comments have all been taken into consideration in this final Consensus Document.

The text of the document, which follows, will make constant reference to the six ITG papers as well as those from the Technology Group. Readers who require further details and full references are advised to consult the original documents in the British Journal of Nutrition[1] and in Trends in Food Science and Technology.[2] It should be emphasized that the intention of this Consensus Document was to select specific examples to highlight some aspects of the subjects covered. It was not the intention for it to be a comprehensive and exhaustive account.

It also must be borne in mind that the subjects of the ITGs were identified on the basis that they represented most, but not all, of the important physiological functions related to this field. The exclusion of subjects in these papers should not be taken to indicate that they are unimportant. Because scientific knowledge is continually advancing, the Consensus Document cannot pretend to be totally up to date. Nevertheless, it has been based on the most recent data available at the time of writing (published up to the end of 1997).

1.4 Structure of the Consensus Document

After the introductory Section of this Consensus Document, Section 3 will outline the aims of functional food science and will introduce some important considerations about classification and criteria for markers of target functions, which play a major role in health.

Section 4 will consider each of the areas of physiological function reviewed in the six ITG papers and will highlight some of the key target functions with examples of relevant markers wherever possible. Also included in the consideration are the options for application (*i.e.* how the target functions could be modulated positively by components within functional foods), as well as proposals for research opportunities.

Section 5 of this document will consider how food technology can play an important role in the development of functional foods and will draw on examples of key technological challenges from the five areas covered in the technology theme papers. Research opportunities that could help to meet the challenges will be indicated.

Finally, Section 6 of this Consensus Document deals with the communication of health benefits to the public and the principles, definitions, use and scientific basis of claims.

Although this area is primarily a regulatory concern and not within the mandate of ILSI as a scientific organization, we recognize that communication of health benefits is an essential element in improving public health in which science plays a vital role. This section offers a brief overview, using examples of types of claims and of local regulatory philosophies and approaches. The aim is to explain/highlight the differences in claims and approaches based on nutritional, physiological or pathophysiological scientific knowledge. This EU Concerted Action has allowed the development of ideas that suggest an innovative way in which to link the science of functional foods with the communication about their possible benefits to consumers.

The Executive Summary, which includes the Recommendations of the Consensus Document, is found in Section 6, and the Key Messages are summarized in Section 7.

The following scientists reviewed and agreed on the text of this Consensus Document:

Co-ordinator:
Dr Berry Danse, ILSI Europe, 83 Avenue E. Mounier, Box 6, B-1200 Brussels, Belgium.
Scientific co-ordinator:
Prof. Marcel Roberfroid, Catholic University of Louvain, Ecole de Pharmacie, Tour Van Helmont, 73 Avenue E. Mounier, B-1200 Brussels, Belgium.
EC responsible:
Dr Liam Breslin, Agro-industrial Research, Food, Commission of the European Communities, Directorate-General XII, Science, Research and Development, 200 Rue de la Loi, B-1049 Brussels, Belgium.
Project manager:
Dr Laura Contor (for further information), ILSI Europe, 83 Avenue E. Mounier, Box 6, B-1200 Brussels, Belgium. Tel. +32-2 771.00.14 Fax +32-2 762.00.44

1.5 About ILSI

The International Life Sciences Institute (ILSI) is a non-profit, world-wide foundation established in 1978 to advance the understanding of scientific issues related to nutrition, food safety, toxicology and the environment. By bringing together scientists from academia, government, industry and the public sector, ILSI seeks a balanced approach to solving problems with broad implications for the benefit of the general public. ILSI is affiliated with the World Health Organization as a non-governmental organization and has specialized consultative status with the Food and Agriculture Organization of the United Nations. Headquartered in Washington, DC, USA, ILSI has branches in Argentina, Australasia, Brazil, Europe, India, Japan, Korea, Mexico, North Africa and the Gulf Region, North America, South Africa, the South Andean region, Southeast Asia and Thailand, and a Focal Point in China. Today, ILSI enjoys the support of around 300 companies and a network of scientists throughout the world.

1.6 About ILSI Europe

In 1986, ILSI Europe was created to focus on the specific needs defined by the Institute's European partners.

The main goals of ILSI Europe are to:

- foster scientific advances by promoting collaboration among scientific experts in industry, academia, and national and international regulatory bodies;
- provide coherent scientific answers to scientific issues of common concern for the well-being of the general public;
- support an active publication programme for the dissemination of scientific information to the broadest possible audience including the scientific community, international organizations and regulatory agencies.

To address these issues, ILSI Europe's members initiate projects, which are managed by specific task forces. Task forces accomplish their goals through activities such as research, workshops, conferences and publications.

1.7 About the European Commission FAIR RTD programme

This concerted action 'Functional Food Science in Europe' (FUFOSE) has been funded within the FAIR RTD programme, which is part of the Commission's Fourth Framework Programme for research and technological development.

This programme aims at promoting trans-European research in the primary production sectors of agriculture, horticulture, forestry, fisheries and aquaculture, linking these with the input and processing industries, particularly food processing and renewable biomaterials.

The food area is important within this programme and is covered by the theme 'Generic Science and Advanced Technologies for Nutritious Foods'.

There is growing interest in Europe in the concept of 'Functional Foods' and this Concerted Action, bringing together Europe's scientists and industry, is fundamental to establishing a science-based approach to such foods.

2 Introduction

2.1 From traditional to new concepts in nutrition

The primary role of diet is to provide sufficient nutrients to meet the metabolic requirements of an individual and to give the consumer a feeling of satisfaction and well-being through hedonistic attributes such as taste. However, in addition to this there is evidence to support the hypothesis that, by modulating specific target functions in the body, diet can have beneficial physiological and psychological effects beyond the widely accepted nutritional effects. In fact, diet can not only help to achieve optimal health and development, but it might also play an important role in reducing the risk of disease.

We are at a new frontier in nutrition science because, at least in the industrialized world, concepts in nutrition are changing significantly. We are progressing from a concept of 'adequate nutrition' to one of 'optimal nutrition'. We have moved from a former emphasis on survival, through one of hunger satisfaction and of food safety, to our present emphasis on the potential for foods to promote health, in terms of both improving well-being (mental and physical conditioning) and reducing the risk of diseases.

Although there are still many people who know little nutrition science, consumer awareness of the subject and its relationship to health is, nevertheless, growing appreciably. Almost everyone today is more conscious of and better informed about the subject than they were in the past. As a result, their expectations of obtaining health benefits from the food they consume is increasing. In a recent survey of 14 331 people interviewed in all 15 Member States of the European Union, 9 % chose 'eating healthily' as the most important influence in selecting food and 32 % said that this had some influence of their food choice.[3]

2.2 From the Importance of Improving Life Expectancy to the Importance of Quality of Life

The changing concepts in nutrition are of particular importance in view of some significant trends in our present society. They are:

1. The increasing cost of health care and of days lost from work.
2. The continuing increase in life expectancy.
3. The increase in the numbers of elderly people.
4. The desire of people for an improved quality of life.

The rising cost of health care is of primary concern in many parts of the world. The cost of health care in European countries in 1995 was estimated to be, on average, around 8 % of the gross domestic product, approximately one percentage point higher than it was a decade ago.

Life expectancy in virtually every country of the world is higher today than it has ever been for all age groups -- from birth as well as at later ages. Expectation of life from birth for both men and women is highest in Japan, but is followed closely by a number of European countries. Furthermore, the rate of increase in life expectancy appears to show no sign of declining.

Associated with this increase in life expectancy is the growth in the size of the population above the age of 65 years. In Europe the relative proportion is currently between 14–17 % of the total population and several European countries have the highest population of elderly people in the world. This figure is expected to rise to between 20–24 % in the next 30 years.

An improved quality of life must accompany this improved life expectancy and increase in the elderly population if relative health care costs can ever be better controlled and managed.

2.3 From New Concepts in Nutrition to Functional Foods

The wide range of food products available to today's consumer offers a wide variety of complex food components, both nutritive and non-nutritive. These have the potential to improve the health and well-being of individuals and, maybe, to reduce the risk from, or delay the development of, major diseases such as cardiovascular disease (CVD), cancer and osteoporosis.

Advances in food science and technology are now providing the food industry with increasingly sophisticated methods to control and alter the physical structure and the chemical composition of food products. There is now a growing realization of the market potential for functional foods, based on the principle of added value linked to health benefit.

2.4 International Developments Relating to Functional Foods

The new concepts in nutrition outlined in Section 2.1 have, over the last 10–12 years, justified the efforts of health authorities in many countries, especially in Japan and in the United States of America, to stimulate and support research on physiological effects of food components and their health benefits. They have also been a major factor in the reconsideration of their regulatory policy on foods and health claims.

In Japan, research on functional foods began in the early 1980s, when 86 specified programmes on 'Systematic analysis and development of food functions' were funded by the government. Later, the Ministry of Education sponsored additional focal point studies on 'Analysis of physiological regulation function of food' and 'Analysis of functional foods and molecular design'. Then, in 1991, the concept of Foods for Specified Health Use (FOSHU) was established. These foods are included as one of the four categories of foods, described in the 'Nutrition Improvement Law' as 'Foods for special dietary use' (*i.e.* 'foods that are used to improve people's health and for which specific health effects are allowed to be displayed'). Upon satisfactory submission of comprehensive data documenting the scientific evidence in support of a proposed health claim, the Minister of Health and Welfare is able to approve a claim, and grant permission, to use a 'symbol' on labelling, to indicate to the consumer that the health claim has government approval.

Foods identified as FOSHU are required to provide evidence that the final food product is expected to exert a health or physiological effect; data on the effects of isolated individual components are not sufficient. FOSHU products should be in the form of ordinary foods (*i.e.* not as pills or capsules) and are assumed to be consumed as part of an ordinary diet (*i.e.* not as very occasional items linked to specific symptoms). Most FOSHU products currently approved contain either oligosaccharides or lactic acid bacteria for promoting intestinal health.

In the United States of America, 'reduction of disease risk' claims have been allowed since 1993 on certain foods. These contain components where the Food and Drug Administration (FDA) has accepted there is objective evidence for a correlation between nutrients or foods in the diet and certain diseases on the basis of 'the totality of publicly available scientific evidence, and where there is substantial agreement amongst qualified experts that the claims were supported by the evidence'. By 1998 there were eleven FDA-approved correlations between foods, or components, and diseases. These are claims for the relationship between foods that are high in calcium and the reduced risk of osteoporosis; claims for foods that are low in saturated fats, low in cholesterol and low in fat and the reduced risk of coronary heart disease; and the claim for sugar alcohols in relation to reduced risk of dental caries. The claim relating diets containing soluble fibre with the reduced risk of coronary heart disease has been amended twice to allow claims for the soluble fibre from whole oats and from psyllium seed husk. Very recently, the FDA has announced that claims can also be based on 'authoritative statements' of a Federal Scientific Body, such as the National Institute of Health and Center for Disease Control, as well as from the National Academy of Sciences, as allowed by the FDA Modernization Act of 1997.

In the European Union, there is no harmonized legislation on health claims, which means that they are dealt with at a national level. However, it is well recognized that the competitive position of the European food and drink industry should be reinforced through a better understanding of the scientific basis for the functionality of food.

A self-regulating programme on health claims was introduced in Sweden in 1990 and revised in 1996. It permits claims with two parts: information on one of eight approved diet–health relationships, followed by information on the composition of the product (function-based claims). The accepted conditions are obesity (energy), blood cholesterol (fat quality), blood pressure (sodium), atherosclerosis (blood pressure, serum cholesterol, long-chain $n-3$ polyunsaturated fatty acids (PUFA) in fish), constipation (dietary fibre), osteoporosis (calcium), dental caries (easily fermentable carbohydrates) and iron deficiency (iron). The food industry and retail organizations behind the programme have recently suggested an extension to cover product-specific physiological claims, characteristic of functional foods.

2.5 From Functional Foods to Functional Food Science

Up to now, the approaches used for functional food science, both in Japan and to a lesser extent in the USA, to match these new concepts in nutrition have mostly been 'product or food component-driven', and they are likely to be very much influenced by local, traditional or cultural characteristics. A science-based, 'function-driven' approach is preferable, because the functions and their modulation are universal. Functional food science, therefore, refers to the new concepts in the science of nutrition that lead to the stimulation of research and to the development of functional foods.

2.5.1 Working definitions. No universally accepted definition for functional foods exists. In fact, because functional foods are more of a concept than a well-defined group of food products, a working definition rather than a firm definition is preferred for the purposes of this Consensus Document.

A food can be regarded as 'functional' if it is satisfactorily demonstrated to affect beneficially one or more target functions in the body, beyond adequate nutritional effects, in a way that is relevant to either an improved state of health and well-being and/or reduction of risk of disease. Functional foods must remain foods and they must demonstrate their effects in amounts that can normally be expected to be consumed in the diet: they are not pills or capsules, but part of a normal food pattern.

A functional food can be a natural food, a food to which a component has been added, or a food from which a component has been removed by technological or biotechnological means. It can also be a food where the nature of one or more components has been modified, or a food in which the bioavailability of one or more components has been modified, or any combination of these possibilities. A functional food might be functional for all members of a population or for particular groups of the population, which might be defined, for example, by age or by genetic constitution.

2.5.2 The functional food science approach for Europe. The design and development of functional foods is a key issue, as well as a scientific challenge, which should rely on basic scientific knowledge relevant to target functions and their possible modulation by food components. Functional foods themselves are not universal, and a food-based approach would have to be influenced by local considerations. In contrast, a science-based approach to functional food is universal and, because of this, is very suitable for a pan-European approach. The function-driven approach has the science base as its foundation – in order to gain a broader understanding of the interactions between diet and health. Emphasis is then put on the importance of the effects of food components on well-identified and well-characterized target functions in the body that are relevant to health issues, rather than solely on reduction of disease risk.

3 The Scientific Basis of Functional Foods

3.1 The Aims of Functional Food Science

Knowledge of the mechanism(s) by which functional food(s) can modulate the target function(s) and their relevance to the state of well-being and health and/ or reduction of a disease will originate from basic knowledge in the biological sciences. It might also be supported by epidemiological data that could demonstrate a statistically validated and relevant relationship between the intake of individually specified food components (better still, a serum, faecal, urinary or tissue marker of intake of the components under study) and the specific benefit. Furthermore, it will be of particular value to have good prospective evidence that links the habitual intake of specified food components with the reduction of the risk of disease, which might develop some time later.

The aims of functional food science are, therefore, as follows:

- To identify beneficial interaction(s) between a functional component within a food and one or more target functions in the body and to obtain evidence for the mechanism(s) of these interactions. (Results from studies carried out *in vitro*, in cells in culture, the use of *in vitro/ex vivo* models and animal models, as well as results from human studies, should be included.)
- To identify and validate markers relevant to these functions and their modulation by food components (see below).
- To assess the safety of the amount of food or its component(s) needed for functionality. This will require evidence that is equally applicable to all major groups in the population, including those who are indulging in behaviour that might be expected to compromise the anticipated benefits of the functional food. It might involve post-marketing monitoring, including the effects on the whole diet.
- To formulate hypotheses to be tested in human intervention trials that aim to show that the relevant intake of specified food components is associated with improvement in one or more target functions, either directly, or in terms of a valid marker of an improved state of health and well-being and/or a reduced risk of a disease.

3.2 Target Functions for Health Outcomes

This Consensus Document on 'functional food science' does not intend to summarize the science and the technology theme papers. Rather, its aim is to illustrate the concepts presented above, by selecting from each theme paper a few topics based on the identification of key target functions:

- that play a major role in maintaining an improved state of health and well-being and/or reduction of risk of disease;
- for which appropriate markers are available and/or feasible;
- for which potential opportunities exist for modulation by candidate food components.

3.3 Markers – A Proposal for Classification

One key, but difficult, approach to the development of functional foods is the identification and validation of relevant markers that can predict potential benefits or risks relating to a target function in the body.

If markers represent an event directly (*i.e.* causally) involved in the process they should be considered as factors, whereas if they represent correlated events they should be considered as indicators.

Markers relevant to functional foods (see Figure 1) can also be classified according to whether they:

- relate to the exposure to the food component under study, such as a serum, faecal, urinary or tissue marker. For instance, the increased level of red-blood-cell folate is a marker of exposure to folate in food and the increased level of blood tryptophan is a marker of exposure to tryptophan in food. Markers relating to exposure to the functional food component can give some indication, but not absolute proof, of the bioavailability of the food component.
- relate to the target function or biological response, such as changes in body fluid levels of a metabolite, protein or enzyme (*e.g.* the reduction in levels of plasma homocysteine as a possible response to dietary folate, and the increased levels of brain serotonin as a possible response to dietary tryptophan).
- relate to an appropriate intermediate endpoint of an improved state of health and well-being and/or reduction of risk of disease, such as the measurement of a biological process that relates directly to the endpoint (*e.g.* the extent of narrowing of the carotid artery as evidence of cardio-vascular disease, or functional imaging of the brain by magnetic resonance imaging as an intermediate endpoint marker for the amelioration of depression).

Markers of exposure and markers of biological response can either be factors that are causally related to the endpoint, or indicators that are indirectly related. Markers of an intermediate endpoint are likely to be factors, rather than indicators.

Markers become less specific and more attenuated and subject to confounding variables as they become more remote from the endpoint. Conversely, they become more specific and quantitatively related the closer they are to the endpoint in question. The elucidation of the mechanisms leading to health outcomes would refine the identification of markers and their appreciation.

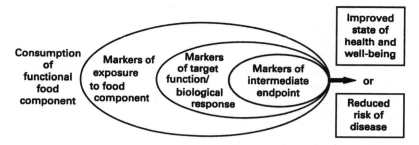

Figure 1 *Classification of markers relevant to the effects of functional foods. This is a diagrammatic representation to show how different types of markers would be expected to lie within a logical progression from the food component to the health outcome. The types of markers are completely independent of each other. Markers can be either indicators, or, if they can be proven to be causal, factors (see Section 2)*

This differential classification is considered to be of real importance in the development of new markers for use in human studies. The results from such studies can also form a scientific basis for formulating and controlling claims (see Section 6).

3.4 Criteria for Markers

In general, all markers, whether they are biochemical, physiological or behavioural in nature, should be feasible, valid, reproducible, sensitive and specific. The following criteria should apply equally well, for example, to measurements of particular blood components as they do to measurements of subjective experience and behaviour.

- Markers should represent relatively immediate outcomes, which can be used to assess interventions in a reasonable timescale; they could, therefore, wherever possible, replace later and more remote outcomes as have been used in some epidemiological studies.
- Markers must be rigorously validated and amenable to standard quality-control procedures.
- Markers must be clearly linked to the phenomena involved in the biological process being studied. It is important to prevent the pursuit of increasingly accurate and precise measurements, which have limited biological significance.
- Markers should undergo single-centre studies to establish their sensitivity (*i.e.* the frequency of a negative test when the process is present) and their specificity (*i.e.* the frequency of a positive test when the process is absent). They must also be shown to be reproducible in different centres.
- Markers must be measurable in easily accessible material, or obtainable using methodology that must be both ethical and minimally invasive.
- Dynamic responses might be as useful as, or more useful than, static

measurements. For example, changes in markers during clearance studies and in postprandial situations and studies of enzyme function, induction and suppression should be considered.

- Appropriate static and dynamic markers might also be based on objective assessments of psychological and physical performance and subjective assessments of quality of life or other similar outcomes.

These criteria for markers should be taken into account before human intervention studies are carried out. In many cases a 'battery' of markers might be needed in order to create a 'decision tree' from multiple tests. A new generation of human intervention studies using markers that accord with these criteria will generate readily interpretable, valid and reliable data, which can form the basis for future development of functional foods in the European population. Markers, arising from new techniques such as molecular biology, might be expected to identify target groups who could benefit from functional foods.

3.5 Safety Considerations

Functional foods must be safe according to all standards of assessing food risk. However, the concept of risk versus benefit cannot be applied in such a straightforward manner as it is for drugs; new concepts and new procedures will need to be elaborated and validated.

For example, the safety evaluation of micronutrients must take into account potential adverse effects of low intakes (clinical deficiency) as well as effects from intakes that are too high (clinical toxicity). Furthermore, a more traditional toxicology testing approach might be suitable to assess the safety of (phyto)chemicals, the daily intake of which remains low, but it is not necessarily ideal for new functional food components, which might account for a relatively larger percentage of the total food intake. The classical 'dose–effect relationship' might lead to considerations of physiological/nutritional disturbances that are irrelevant to standard safety assessment. If a functional food is considered in the framework of novel foods, the decision tree leading to a relevant package of toxicity testing will be driven by the principle of substantial equivalence of the tested food compared with traditional counterparts.[4]

Protocols for human nutrition studies need to be developed including, in some cases, post-marketing surveillance. Even though the design of clinical studies as used in drug development can serve as a reference point, specific protocols and specific criteria relevant to functional foods might be needed. It might also be necessary to identify specific target groups of individuals who might present higher/lower susceptibilities to potential adverse effects and to consider that the effects of functional foods might be positive in some target groups and negative in others. Finally, the long-term consequences of the interaction(s) between functional food components and function(s) in the body and the interactions between components must be carefully monitored. This EU Concerted Action proposes that markers as described in Sections 3.3 and 3.4 should, if possible, be used and integrated in the safety assessment.

4 Target Functions and Research Opportunities in Relation to Health Outcomes (see reference 1, pages S5–S45)

4.1 Growth, Development and Differentiation

4.1.1 Introduction. Food supply, and the metabolism of food components, during both pregnancy and lactation, have important implications on intra-uterine and postnatal development of children. Early nutrition might modulate growth and development of the organism, which is thought to affect neural function and behaviour. Equally important, the effects of early nutrition also appear to influence the overall quality of life by exerting life-long programming effects on modulating health, disease and mortality risks in adulthood.

4.1.2 Key target functions
4.1.2.1 Maternal adaptations during pregnancy and lactation. The energy cost of development of the placenta and fetus is only about 6 % of the mother's normal requirements. This value is relatively low and is probably due to adaptations during pregnancy, which enable the mother to increase the absorption and use of nutrients from her diet, and also the mother's ability to use stored nutrients for fetal growth. Some of these adaptations occur very early in pregnancy; others, such as those involving the increased absorption of iron in the gut, take place in the later stages. Some adaptations seem to target specific nutrients such as iron, towards the placenta for transfer to the fetus, but these transport processes have not been well characterized.

It is important to ensure that mothers are adequately nourished during lactation and between pregnancies with every nutrient, so as to 're-stock' any nutrient storage depots that might have been depleted during pregnancy.

4.1.2.2 Fetal development. Fetal development is under genetic control, which can be disturbed by severe deprivations or excesses of nutrients. The critical periods of fetal organ formation often occur before the fetus has a placental supply of nutrients from the mother. The best documented example of this is the spectrum of iodine deficiency disease. Poor iodine status is associated with an increased risk of miscarriage in early gestation and preterm delivery, as well as abnormal mental development of the infant. However, the mechanisms and regulation of supplying nutrients to the early embryo and their susceptibility to diet are not clear.

Another more recent example is the critical period for the effect of additional dietary intakes of folate on the optimal development of the neural tube – a very early component of the nervous system. Intervention at this critical stage can result in a reduction in incidence of a number of neural tube defects, one of which is spina bifida.

4.1.2.3 Infant and child growth and development. Postnatal growth and development of the infant is a continuation of the *in utero* events and processes. Although the nature and means of delivering nutrients has changed significantly, the functional and structural integrity of organs and tissues such as the gut, liver, skeleton, brain, adipose tissue, muscle, blood cells and the immune system continue to mature and remain potentially susceptible to dietary factors.

The newborn infant has an intermittent nutrient supply from breast milk or infant formula, both of which are rich in lipids. Absorption of nutrients takes place through the intestine, and delivery to the tissues is initially *via* the portal circulation. At the transition between suckling and weaning, there is a switch to a more solid diet with a lower fat and higher carbohydrate content.

The adaptation of infants to these changes requires important modifications of substrate and energy metabolism through the induction of metabolic pathways. The programming of metabolic responses at the cellular level, the programming of stem cells and the control of programmed cell death are not well understood. Postnatal development and differentiation also creates distinctive requirements for nutrients such as amino acids and long-chain fatty acids. Some of these, although not regarded as essential in adults, are considered to be 'conditionally' essential in the infant.

There is now epidemiological evidence that early diet, and presumably related metabolic events *in utero* and in infancy, manifest themselves by inappropriate intra-uterine and postnatal growth. This is now thought to predispose the child to obesity, diabetes mellitus, and risk factors for CVD in adult life. These later consequences might arise from alterations in critical stages in metabolic programming of several nutrients and might be analogous to the early consequences of deficiency of iodine and folate.

Breast milk contains some constituents derived directly from the maternal diet that might not only have an obvious role to play in determining the baby's perception of smell and taste (which is well developed *in utero*) but also in the development of immunotolerance. Thus it is possible that early diet might condition the development of taste preferences, which, in turn, could affect the diversification of the diet in later infancy.

4.1.3 Examples of options for modulation. Human breast milk is an example of a food with functional properties. The precise roles and essentiality of many of the diverse components of breast milk are not certain and further clarification of their role might be informative for the development of functional foods. Bioactive components in human milk include the specific nature of the protein and lipid composition, digestive enzymes, growth factors, peptide and steroid hormones, free amino acids (taurine, arginine, glutamine), choline, polyamines, nucleotides, non-protein nitrogen sources such as urea, immunoglobulins and oligosaccharides.

These components in breast milk might be present adventitiously or they

Table 1 Examples of opportunities for modulation of target functions related to growth, development and differentiation by candidate food components with possible markers*

Target functions	Possible markers	Candidate food components
Maternal adaptation during pregnancy and lactation	maternal weight body fat infant birth weight milk volume and quality	micronutrients n−3 and n−6 PUFA energy
Skeletal development	ultrasound measures anthropometric measures bone mineral density (e.g. DEXA)	calcium vitamin D vitamin C
Neural tube development	ultrasound measures	folic acid
Growth and body composition	anthropometry body fat mass total body water procollagen propeptide excretion urinary creatinine excretion	growth factors essential amino acids unsaturated fatty acids
Immune function	cellular and non-cellular immune markers	vitamin A vitamin D antioxidant vitamins n−3 and n−6 PUFA trace elements arginine nucleotides and nucleosides probiotics
Psychomotor and cognitive development	tests of development, behaviour, cognitive function and visual acuity (electro-) physiological measurements	n− and n−6 PUFA iron zinc iodine

* The information given in this table is derived from the Theme Paper (see Bellisle *et al.* 1998, S5–S45), where, if not given in this Consensus Document, further details of options for modulations, markets and safety issues can be found.   n−6 PUFA = n−6 series long-chain polyunsaturated fatty acids.
n−3 PUFA = n−3 series long-chain polyunsaturated fatty acids.
DEXA = dual-energy X-ray absorptiometry.

might be present for specific purposes, which have been developed under evolutionary pressure. In any case, study of their function can possibly inform the development of infant formulae and might have wider implications.

4.1.4 Research opportunities
4.1.4.1 Basic science

- The development of the interaction between food components and gene expression needs to be better elucidated. Targeting of genes affected by, and the nature of interaction with, specific nutrients such as glucose, cholesterol, fatty acids and amino acids needs to be characterized.
- To what extent is metabolic maturation intrinsically controlled? How much is it responsive and determined by external stimuli such as the pattern of feeding and changes in the nature of the diet?
- The bioactive compounds in breast milk should be characterized in further detail. There might be opportunities to isolate and include them as functional components in infant formulae or other foods.
- Research is required on the interactions of different dietary components on the development and long-term function and integrity of intestinal function (substrate transporters, mucosal immunity and xenobiotic metabolizing systems), intestinal growth and microflora.
- Oral immunotolerance and dietary factors modulating the development of food protein allergy need better characterization.

4.1.4.2 Cross-sectional and prospective epidemiological evidence

- What is the interaction between the ingestion of human breast milk and the programming and development of infant metabolism during the subsequent introduction of diversified diet?
- Possible mechanisms must be identified to explain the epidemiological associations between impaired intra-uterine and postnatal growth and the prevalence, in adulthood, of obesity, diabetes mellitus, hypertension, hypercholesterolaemia, cardiovascular disease, obesity and other diseases. Well-structured prospective studies, using appropriate markers, are needed.

4.1.4.3 Markers

- There is a general need to apply many of the markers developed for the study of intermediary metabolism and well-being (see later in this Section) to the prospective study of the long-term impact of feeding in early childhood.

4.1.4.4 Human intervention studies

- The nutritional and metabolic basis of the benefits of breast feeding and of human milk need better understanding. What are the maternal dietary factors that influence the composition of human milk? The early differences in substrate metabolism, body composition, the efficiency of growth and functional development between breast-fed babies and those fed infant formula might be important.

4.1.4.5 Evidence of safety

- We need the scientific basis to be more specific about nutritional needs during weaning and childhood to be able to address the potential risks of infants and children being given an adult diet too early or too late.
- ► Weaning and the adaptation to an adult diet is assumed to be a gradual process, but this has not been systematically established.
- Although all of childhood involves growth and development, the most critical periods are probably those *in utero* and during early infancy, as well as the growth spurts during pre-school age and during puberty. The development of functional foods for pregnant women and lactating mothers will need special consideration of the possible adverse effects of these foods on the fetus and child.

4.2 Substrate Metabolism (see reference 1, pages S47–S75)

4.2.1 Introduction. A number of chronic diseases such as obesity, non-insulin-dependent diabetes mellitus (NIDDM) and osteoporosis are partly related to changes in total food intake, levels of physical activity and a poorly balanced diet.

The balance of the diet can determine substrate metabolism and the optimally balanced diet is usually expressed in terms of its macronutrient content. The carbohydrates are subdivided chemically into monosaccharides, disaccharides, oligosaccharides, starches, and non-starch polysaccharides, polyols and alcohol. Nutritionally, a most important distinction is between carbohydrates that are digested and absorbed in the small intestine and those that are not. The lipids are defined according to their fatty acid composition, particularly the relative contents of saturated fatty acids (SFA), mono-unsaturated fatty acids (MUFA) (cis and trans) and the polyunsaturated fatty acids (PUFA), which can be further sub-divided into the PUFA of the $n-6$ series and the PUFA of the $n-3$ series.

A number of adverse metabolic changes and other cardiovascular risk factors tend to 'cluster' within individuals. These include high blood pressure, high plasma levels of insulin and/or glucose, high plasma levels of triacylglycerols (TAG) in the fasting state and after a meal, a preponderance of small, dense low-density lipoprotein (LDL) particles and low levels of high-density lipoprotein cholesterol (HDL). The terms 'Syndrome X', 'metabolic syndrome' or 'insulin-resistance syndrome' have been used to describe the whole cluster of cardiovascular risk factors since the underlying change appears to be an increase in insulin resistance (or a decrease in insulin sensitivity) of the tissues. The characteristic changes in the lipid profile have been called the Atherogenic Lipoprotein Phenotype (ALP). Each of these factors has been shown to relate independently to the risk of CVD, and insulin resistance is a strong marker of the risk development of NIDDM.

During physical stress such as exercise, the substrate demands are enormous and a balanced diet with a carefully planned mix of food components can play a crucial role in improving the level of performance.

4.2.2 Key target functions

4.2.2.1 Maintenance of appropriate body weight, body composition and body fat distribution. Obesity develops when the energy intake is consistently greater than energy expenditure. The excess energy is stored in the form of adipose tissue and results in a body weight and total body fat content that is excessive for height.

Obesity, particularly a tendency to store fat in internal, visceral depots ('central obesity'), is associated with an increased risk of developing high blood pressure, insulin resistance, diabetes and CVD. Since obesity and relative fat distribution are themselves, in part, genetically determined, there might be a considerable genetic component to the insulin resistance syndrome. Nevertheless, they must also be strongly influenced by environmental factors including diet and physical activity level. There has been a shift towards the consideration of markers that can distinguish 'central obesity' as well as the more traditional body markers for total body fat in the determination of the health risks of obesity.

4.2.2.2 Control of macronutrient oxidation. Achievement of macronutrient balance requires that the net oxidation of each macronutrient equals the amount of that nutrient in the diet. Oxidation of carbohydrates and proteins tend to vary in response to the recent intake of each fuel and thus these two fuels appear to regulate their own oxidation. In contrast, when carbohydrate content of the diet is constant, fat intake, particularly in the short term, does not always directly promote its own oxidation.

4.2.2.3 Regulation of thermogenesis. The increased energy expenditure after a meal (diet-induced thermogenesis) is mediated by a prolonged activation of the sympathetic nervous system leading to an increased catecholamine release.

4.2.2.4 Control of insulin sensitivity and blood glucose control. Insulin sensitivity is usually measured as the ability of insulin to stimulate glucose disposal, usually by means of the 'clamp' technique. The measurement of the fasting plasma insulin and glucose concentrations can be used as simple markers, suitable for large-scale epidemiological studies. Fasting blood glucose values in the upper part of the normal range, which become mildly elevated after a meal (impaired glucose tolerance), might represent a cardiovascular risk factor. Values that exceed the upper limit of the normal range might become the primary cause of the long-term microvascular complications of diabetes (such as those that affect the retina, the kidneys and the nervous system).

4.2.2.5 Control of plasma triacylglycerols. The combination of elevated plasma TAG concentrations and low HDL-cholesterol concentrations is a particularly strong risk marker for CVD. This is the characteristic disturbed lipid profile associated with insulin resistance. The magnitude and duration of elevated TAG concentrations after a meal (postprandial lipidaemia) is a strong marker of CVD risk, even in the presence of normal fasting TAG concentrations.

4.2.2.6 Optimal performance during physical activity. Training and competition will increase the daily energy expenditure by between 500 and 1000 kcal per hour of exercise, depending on its intensity. Large sweat losses might pose

Table 2 Examples of opportunities for modulation of target functions related to substrate metabolism by candidate food components with possible markers*

Target functions	Possible markers	Candidate food components
Maintenance of desirable body weight	BMI, body fat content anthropometric indices and imaging techniques as indirect and direct measures of 'central obesity' respiratory quotient resting metabolic rate	energy density ($\downarrow$) fat replacers carbohydrate: fat ratio food with low glycaemic index fibre
Control of blood glucose levels and insulin sensitivity	fasting glucose postprandial glucose glucose tolerance test glycosylated haemoglobin measures of insulin dynamics fasting plasma insulin	as for body weight control, plus: food with low glycaemic index soluble viscous fibre saturated fatty acids ($\downarrow$)
Control of TAG metabolism	fasting plasma TAG postprandial plasma TAG	as for body weight control, plus: n−3 PUFA n−6 PUFA monounsaturated fatty acids
Optimal performance during physical activity	body temperature performance testing muscle mass muscle protein synthesis	water/electrolytes energy high and low glycaemic index carbohydrates ergogenic substances such as creatine protein/specific amino acids
Fluid homeostasis	water balance electrolyte balance	isotonic carbohydrates electrolyte fluids

* The information given in this table is derived from the Theme Paper (see Bellisle *et al.* 1998, S47–S75), where, if not given in this Consensus Document, further details of options for modulations, markets and safety issues can be found.
($\downarrow$) = reduced intake of candidate food component.   TAG = triacylglycerol
n−3 PUFA = n−3 series long-chain polyunsaturated fatty acids.   n−6 PUFA = n−6 series long-chain polyunsaturated fatty acids.

a risk to health by inducing severe dehydration, impaired blood circulation and heat transfer. This will ultimately lead to heat exhaustion and collapse. Insufficient replacement of carbohydrates might lead to low blood glucose levels, fatigue and exhaustion.

The requirements for specific nutrients and water depends on the type, intensity and duration of the physical effort. Specific nutritional measures and dietary interventions can be devised that are particularly appropriate for the distinct phases of preparation, competition and recuperation.

4.2.3 Examples of options for modulation

- Foods with reduced energy and/or fat content can help to control body weight and to improve glucose tolerance and insulin sensitivity. Further developments will include the use of new fat replacers and fat substitutes and the search for food components that can specifically modulate 'central obesity'. Saturated fatty acids could impair insulin sensitivity thus increasing the risk of developing NIDDM. In contrast, $n-6$ PUFA (but not $n-3$ PUFA) might decrease this risk, but this needs to be confirmed by intervention trials.
- The glycaemic index (GI) can be used to categorize foods and is defined as the incremental blood glucose area after the test product has been ingested. It is expressed as a percentage of the corresponding area after ingesting a glucose load or a carbohydrate equivalent amount of white bread – both of which have high GIs. An increase in foods with a lower GI in the diet, such as pasta or legumes, might improve postprandial blood glucose levels and might help to improve insulin sensitivity in those prediposed to NIDDM.
- Non-digestible, fermentable carbohydrates in the diet (resistant starch, non-starch polysaccharides and oligosaccharides and polyols), and, possibly, some of their fermentation products, could also lower blood glucose levels by a decrease in liver glucose production. Cell wall polysaccharides are able to 'encapsulate' dietary carbohydrates and, in so doing, slow their accessibility and thus their digestibility. Soluble viscous fibres consumed in large amounts have the most effect on postprandial glucose and insulin response. The effect is related to viscosity, which inhibits mixing and diffusion in the intestinal tract and possibly delays gastric emptying.
- Caffeine is a potent thermogenic agent of the group of methylxanthines, but other pungent components from spices, such as ginger, chilli and mustard, are also potential thermogenic agents.

4.2.4 Research opportunities

4.2.4.1 Basic science

- Further research is necessary to understand mechanisms responsible for insulin resistance, in particular in relation to the role of intermediary carbohydrate and fat metabolism in the development of 'central' obesity.
- What is the effect of different types of carbohydrate in the regulation of carbohydrate balance? What are the mechanisms behind the impact of the

ratio of PUFA to SFA in the diet on fat oxidation? We need to know the impact of the different short-chain fatty acids (SCFA), such as acetic and propionic acids, on metabolism in general.

- Are there mechanisms, apart from the activation of the sympathetic nervous system, that can account for the elevation of diet-induced thermogenesis?
- Nutrient mechanisms that have an impact on genetic expression or immunological function should be examined. Modern methods of molecular biology and immunology must be used with a focus on post-translational modifications related to various proteins, diets or indispensable substrates. Interactions between different nutrients are an essential focus of future studies because of alterations caused by the ageing process.

4.2.4.2 Cross-sectional and prospective epidemiological evidence

- Is it possible to make a distinction between the genetic and the environmental components of the ALP? ALP appears to be a secondary consequence of insulin resistance but is often regarded as a genetic condition. How prevalent is ALP in non-obese, non-diabetic subjects?
- Is it reasonable to base our understanding of the effects of dietary manipulation within individuals on evidence from epidemiological studies? An elevated plasma TAG concentration, for example, might be a risk factor for CHD in epidemiological terms, but does it then follow that elevation of plasma TAG by dietary means confers an equivalent risk?

4.2.4.3 Markers

- Achievement of the above goals requires the development of analytical methods (bioassay), the identification of markers and a more precise knowledge of nutrient requirements.

4.2.4.4 Human intervention studies

- What are the long-term effects of macronutrient replacers, in particular fat replacers, on energy and fat balance and on body weight control? Is the consumption of fat-reduced or energy-reduced foods eventually compensated for by an increased total food intake?
- Long-term intervention studies are required to investigate whether insulin resistance can be ameliorated by the manipulation of the carbohydrate:fat ratio in the diet and by the composition of the dietary fatty acids. Shorter-term mechanistic studies should also be carried out within these long-term studies to investigate the regulation and rate of adaptation of carbohydrate and fat balance within the body. Potential gene/nutrient interactions must be considered.
- Low GI foods are a useful tool for dietary management of people with NIDDM, and some can help to reduce plasma cholesterol concentrations and insulin resistance. Their further potential should be explored, particularly in relation to the prevention of NIDDM and cardiovascular diseases.
- It is critically important that long-term (at least 6 months) studies of the effects of low-fat, high-carbohydrate diets on plasma lipid constituents are carried out. Is the elevation of plasma TAG concentration on a high-

carbohydrate diet transient or longer lasting? What is the nature of the elevation of plasma TAG concentrations, *i.e.* what are the size and density of lipoprotein particles containing TAG? What is the impact of the elevation of TAG on other lipid constituents (*e.g.* the density distribution of LDL particles) and on postprandial lipidaemia?

- It is important to know the effects of different types of fatty acids and available carbohydrates, as well as the effects of the carbohydrate:fat ratio, on long-term energy and substrate balance.

4.2.4.5 Evidence of safety

- The safety of substances that modulate energy and glucose metabolism should be evaluated in short- and long-term studies because of the risk of disturbances in the lipid and glucose metabolism.

4.3 Defence against Reactive Oxidative Species (see reference 1, pages S77–S112)

4.3.1 Introduction. The normal development and functioning of any aerobic organism is associated with the generation of reactive oxidative species (ROS), referred to as oxidants or prooxidants. ROS are responsible for oxidative damage to biological macromolecules such as DNA, lipids or proteins, which might be involved in the initiation or progression of diseases, such as some forms of cancer and cataract, cardiovascular disease, age-related macular degeneration, rheumatoid arthritis and a number of neurodegenerative conditions such as Parkinson's disease or Alzheimer's disease. Gene regulation and the immune system might also be affected. It is also possible that ROS are involved in intracellular signal transduction.

ROS can arise endogenously as secondary products from normal metabolic reactions, or they can be primarily formed as part of the body's defence against foreign organisms. The body is also exposed to ROS from external sources; many prooxidants are present in the diet and are inhaled in cigarette smoke and in other forms of air pollution. The most relevant ROS are peroxyl radicals ($ROO\cdot$), the nitric oxide radical ($NO\cdot$), the superoxide anion radical ($O_2\cdot-$), the hydroxyl radical ($OH\cdot$), singlet oxygen ($^1O_2$), peroxynitrite ($ONOO-$), and hydrogen peroxide (H_2O_2).

To counteract the oxidant load, a diversity of defence mechanisms operate in biological systems, which include both enzymic and non-enzymic antioxidant systems. An antioxidant has been defined as 'any substance that, when present in low concentrations compared to that of an oxidizable substrate, significantly delays or inhibits the oxidation of that substrate'.

4.3.2 Key target functions

4.3.2.1 Preservation of structural and functional activity of DNA. Oxidative damage to DNA can lead to strand breaks and/or the modification of DNA bases, which might result in point mutations, deletions, or gene amplification. The COMET assay can measure distortion of morphology due to damaged DNA. High-performance liquid chromatography (HPLC) and gas chromatography–mass spectrophotometry (GC–MS) methods can be used to measure damaged DNA bases such as 8-hydroxydeoxyguanine; antibody techniques are currently also being developed for this purpose. However, measurement of gross damage to DNA might give little information about oxidative damage to specific genes whose expression has a key cellular function.

4.3.2.2 Preservation of structural and functional activity of circulating lipoproteins and of polyunsaturated fatty acids in cell membranes. Oxidation of the PUFAs in LDL within the arterial wall plays a major role in the pathogenesis of fatty streak formation in arteries, which, in turn, leads to atherosclerosis. Oxidative modification of both the lipid and protein part of the lipoprotein contribute to the atherogenic properties of oxidized LDL (LDLOX). The products of lipid oxidation are best measured in plasma as lipid hydroperoxides or their derivatives such as malondialdehyde, or as isoprostanes. Development of isoprostane measurement will advance specificity and precision and is expected to provide a marker for whole body lipid peroxidation. Direct measurements of oxidized LDL are also now possible.

Oxidative stress might play a role in the development of neuronal degeneration, since ROS can cause peroxidation of the PUFA in all cell membranes including nerve cells with subsequent membrane disruption.

4.3.2.3 Preservation of structural and functional activity of proteins. Proteins in the lens of the eye are extremely long-lived and often show oxidative damage by ROS through chronic exposure to light and oxygen. Damaged proteins can aggregate and precipitate, giving rise to cataracts and to macular degeneration in elderly people. The methodology for the measurement of oxidative damage to the protein part of lipoproteins and other proteins has not, as yet, been sufficiently developed and evaluated.

4.3.3 Examples of options for modulation

- The human diet contains an array of different compounds that possess either antioxidant activities or the apparent ability to scavenge ROS. Epidemiological studies support the hypothesis that the major antioxidant nutrients vitamin E and vitamin C (as well as β-carotene, which might or might not be acting as an antioxidant *in vivo*) play a beneficial role in the maintenance of health and in the reduction of chronic disorders. LDL oxidation is efficiently inhibited by lipophilic antioxidants, of which α-tocopherol is the most important in *in vitro* systems; β-carotene is, however, ineffective. Carotenoids and polyphenolic compounds such as flavonoids are also potent antioxidants in other *in vitro* systems, but there is no evidence that they function this way *in vivo*. Vitamin C has been demonstrated to be the most important antioxidant in the human eye.

- The food industry has a long experience in the control of oxidative damage in foods and this expertise can be used to advantage for the protection of antioxidants specifically used as food components to be beneficial to health.

Table 3 Examples of opportunities for modulation of target function by candidate food components and possible markers related to defence against reactive oxidative species*

Target functions	Possible markers	Candidate food components
Preservation of structural and functional activity of DNA	measurement of damaged DNA components	vitamin E vitamin C carotenoids polyphenols including flavonoids
Preservation of structural and functional activity of polyunsaturated fatty acids	measurement of lipid hydroperoxides or derivatives	vitamin E vitamin C carotenoids polyphenols including flavonoids
Preservation of structural and functional activity of lipoproteins	measurement of lipid hydroperoxides and oxidized apoproteins	vitamin E vitamin C carotenoids polyphenols including flavonoids
Preservation of structural and functional activity of proteins	measurement of damaged proteins or components	vitamin E vitamin C selenium carotenoids polyphenols including flavonoids

* The information given in this table is derived from the Theme Paper (see Bellisle *et al.* 1998, S77–S112), where, if not given in this Consensus Document, further details of options for modulations, markets and safety issues can be found.

4.3.4 Research opportunities

4.3.4.1 Basic science

- More studies are needed on the uptake, distribution, metabolism and elimination of vitamin E and vitamin C.
- Can flavonoids and carotenoids be taken up by human subjects and distributed to the tissues in quantities sufficient to confer a biological effect, and what is the nature of that effect if it occurs?
- Are there combined effects of synergy or antagonism both with respect to the antioxidants themselves or to other food components?
- Is it the antioxidant role of the substance that is important, or is it some other function, such as some aspects of the modulation of gene expression?
- Further work is needed on the chemical analysis of the antioxidant content of foods so that more realistic food composition tables can be compiled. Ideally this should take account of agricultural practices, industrial processing and food preparation.

4.3.4.2 Cross-sectional and prospective epidemiological evidence

- Better food composition data and better markers of disease could enable re-analysis of existing epidemiological data in some cases and will improve study design in further investigations. Further knowledge of basic science will also allow consideration of confounding influences in epidemiological studies.

4.3.4.3 Markers

- Because direct measurement *in vivo* of the prooxidants that are detrimental to human health is difficult or impossible, it is essential to develop and validate a number of markers of oxidative damage to key body structures that are non-invasive and can be used ethically and acceptably in human volunteers.
- Existing methods for measurement of oxidative damage in human subjects in a non-invasive manner must be refined and validated. Results obtained in the same laboratory on identical material using different, but complementary, methodology should be compared. For the purposes of quality assurance, results from several different participating laboratories of oxidative damage in identical samples must be compared.
- In the case of reduction of disease risk, it is important to show that markers are clearly linked to those phenomena that give rise to a specific human disease since its development is the ultimate paradigm by which the relevance of a marker is judged. For example, which of the multiple oxidative modifications of DNA are those that lead to carcinogenesis? What is the comparative relevance of modification of the protein and lipid in LDL particles or of changes in thickening of the carotid artery wall, all of which contribute to the process of atherogenesis?

4.3.4.4 Human intervention studies

- A new generation of intervention trials using validated and accepted markers as intermediate endpoints might be used to test the efficacy of antioxidants; these will provide results in a shorter time than old-style intervention studies where development of disease was the only marker. Bioavailability studies and dose-response studies, in combination with the development and application of biomarkers, are required for a successful research strategy.

- These new studies will enable quantitation of the optimal levels of intake of antioxidants. Care must be taken to ensure that the importance of the antioxidant contribution of the whole diet, as distinct from that of each individual antioxidant, is evaluated.

4.3.4.5 Evidence of safety

- The mechanism for the apparent exacerbation of the incidence of lung cancer in heavy smokers given supplements of β-carotene needs urgent clarification. Further work is also needed on the safety of other carotenoids, flavonoids and other phenolic compounds that have been shown to have bioactivity in human subjects.
- Developments in food technology will include the preservation, enhancement, and perhaps addition *de novo* to foods, of the particular forms of antioxidants that have been shown to have functional capacity.

4.4 Cardiovascular System (see reference 1, pages S113–S146)

4.4.1 Introduction. CVD is essentially caused by the narrowing of the arteries, which can lead to a reduced supply of oxygen to organs such as the heart, skeletal muscle, brain, intestine and kidneys. Key elements involve changes in the constituents of the blood and in the walls of the blood vessels. The main target functions in this process are those involved with the integrity of the blood vessels (*e.g.* control of cellular immunology and hypertension); in lipoprotein homeostasis (*e.g.* insulin resistance) and the control of the thrombogenic balance (the likelihood of blood clot formation). The interdependence of these factors has not been fully characterized, and since only 50 % of the incidence of CVD can be explained by the known risk factors, there are possibly other unexplored contributory and interactive factors, such as genetic polymorphisms.

4.4.2 Key target functions

4.4.2.1 Lipoprotein homeostasis. A raised plasma concentration of LDL cholesterol is a risk factor for CVD. High levels of lipoprotein(a), very low-density lipoprotein (VLDL), TAG concentrations and low levels of HDL cholesterol are strong risk indicators, rather than risk factors, for CVD.

It is important to consider the lipoprotein profile after a meal (the postprandial response), as well as under fasting conditions, in relation to CVD. For example, the atherogenicity of lipoprotein remnants is more pronounced on high-fat diets, which induce higher postprandial TAG levels, even if fasting TAG levels are low. Long-chain n − 3 PUFA from fish oil might reduce the postprandial TAG response.

The efficacy of dietary manipulations to alter lipoprotein profiles depends on an individual's genetic constitution. For example, there is a genetic polymorphism for one of the apolipoproteins, apolipoprotein E (apoE). People with the apoE2 isoform have a delayed clearance of lipid-rich particles after a meal that is rich in fat, because this type of apolipoprotein has a lower affinity for the receptors on liver cells, resulting in slower removal from the plasma.

4.4.2.2 Endothelial and arterial integrity. Damage to, or activation of, the endothelial cells that line the arteries, as well as more general structural damage at susceptible points in the arteries (such as bifurcations), might increase the risk of atheromatous plaque formation. Endothelial activation increases production, release or exposure of various reactive molecules such as cytokines, adhesion molecules and markers of platelet activation and it causes decreased synthesis of some prostaglandins and transforming growth factors. The consequences might be inflammatory changes and fibrosis of the vascular wall with an increased risk of small blood clot formation. These consequences are all thought to be involved in fibrous plaque formation.

4.4.2.3 Control of thrombogenic potential. The control of thrombogenic potential is likely to be an important element in the reduction of CVD. However, available measures of platelet function *in vitro* and plasma concentration of factors involved in coagulation and in fibrinolysis are not reliable indicators of any pro-thrombotic state or risk in humans. Similarly, the particular influence of endothelial cells on this process, as indicated by relevant markers, has not been sufficiently elucidated.

4.4.2.4 Control of homocysteine levels. Epidemiological data suggest that high plasma levels of homocysteine are associated with increased risk of CVD and several proposed mechanisms for the effects of homocysteine on atherogenesis and thrombosis have been suggested.

4.4.2.5 Control of hypertension. CVD is directly related to systolic and diastolic blood pressure, and antihypertensive measures reduce the risk of coronary disease. Evidence suggests that, in addition to certain food components (*e.g.* sodium, calcium, certain fatty acids), many other factors are involved in the aetiology of hypertension (*e.g.* genetic predisposition, obesity, etc.).

4.4.3 Examples of options for modulation

- Dietary SFA (chain length up to 16 carbon atoms) increase plasma LDL cholesterol concentrations more than they increase plasma HDL concentrations. Stearic acid (C18:0) has less effect on plasma LDL and HDL. Dietary trans-unsaturated fatty acids increase plasma LDL and reduce HDL cholesterol concentrations. Moreover, they increase the plasma lipoprotein(a) concentration. Foods low in SFA and trans fatty acids could, therefore, reduce the risk of CVD.
- The *cis*-unsaturated fatty acids oleic, linoleic and α-linolenic acids reduce plasma concentrations of LDL cholesterol without significantly affecting plasma HDL cholesterol and lipoprotein(a) concentrations. Foods enriched in these unsaturated fatty acids, therefore, could also be used to reduce the risk of CVD.
- Soluble fibre, certain phytosterols and phytostanols and other food

Table 4 Examples of opportunities for modulation of target functions related to the cardiovascular system by candidate food components with possible markers*

Target functions	Possible markers	Candidate food components
Lipoprotein homeostasis	lipoprotein profile, including: plasma LDL-cholesterol plasma HDL-cholesterol plasma TAG (fasting and postprandial response) lipoprotein particle size	#SFA ($\downarrow$) MUFA PUFA certain phytosterols certain phytostanols soluble fibre trans-fatty acids($\downarrow$) soy proteins tocotrienols fat replacers
Arterial integrity	#growth factors adhesion molecules cytokines	certain antioxidants n−3 PUFA from fish
Control of thrombogenic potential	platelet function activated clotting factors and activation peptides	n−3 PUFA from fish certain antioxidants linolec acid
Control of hypertension	systolic and diastolic blood pressure	total energy ($\downarrow$) sodium chloride ($\downarrow$) n−3 PUFA from fish
Control of homocysteine levels	plasma homocysteine levels	folic acid vitamin B6 vitamin B12

* The information given in this table is derived from the Theme Paper (see Bellisle *et al.* 1998, S113–S146), where, if not given in this Consensus Document, further details of options for modulations, markets and safety issues can be found.
($\downarrow$) = reduced intake of candidate food component.   # = see text.   n−3 PUFA = n−3 series long-chain polyunsaturated fatty acids.
SFA = saturated fatty acids   MUFA = monounsaturated fatty acids.   TAG = triacylglycerol   LDL = low-density lipoprotein.
HDL = high-density lipoprotein.

components can reduce LDL cholesterol concentrations, particularly in hyperlipidaemic subjects. Similar effects have been shown for ethanol and for garlic.

- The long-chain n − 3 PUFA found in fish oils can promote mechanisms that improve endothelial and arterial integrity, particularly those dependent on the balance of prostaglandins. They also reduce plasma TAG and may also have suppressive effects on the cellular immune system.
- Diets rich in antioxidants can influence the activities of immune-competent cells and can inhibit the expression of genes coding for the cell–cell adhesion factors.
- There is scope for the modulation of factors such as high plasma homocysteine concentrations and high blood pressure to protect vascular integrity.

4.4.4 Research opportunities
4.4.4.1 Basic science
- Modifications to certain fatty acids, cholesterol and other lipids should be developed in order to influence their absorption and/or systemic handling.
- What is the mechanism of LDL uptake by monocytes and macrophages and what are the factors influencing the access of these cells to the vascular sub-endothelial space?
- The underlying short- and long-term mechanisms, fast-cell signalling-pathway factors involved in cell–cell interactions, thrombogenic and inflammatory reactions should be studied in detail at the level of gene activation and transcription. These could help to explain the links between certain food components and the development and possible regression of vascular lesions.

4.4.4.2 Cross-sectional and prospective epidemiological evidence
- Because cardiovascular disease is a multifactorial condition, we need to know, in the context of nutrition and public health policy, who is likely to benefit from particular dietary/ingredient modifications.
- There is also a need to characterize prospectively, and establish relationships between, population groups and possible risk indicators/factors, such as high blood pressure, high levels of homocysteine in the blood, abnormal lipid profiles, and CVD.

4.4.4.3 Markers
- Prospective validation studies should be performed to establish to what extent the presently available indicators actually reflect the risk of arterial thrombosis. The current available indicators of arterial thrombosis tendency include platelet aggregation *in vitro*, urinary excretion of thromboxane- and prostacyclin metabolites and of specific platelet proteins, plasma concentrations of soluble forms of cell adhesion molecules, activation fragments of clotting factors, and fibrin degradation products.
- It might be necessary to develop and validate new methods to measure *in vivo* arterial thrombosis tendency in humans and to look for, and prospectively validate, more specific *in vivo* activation markers for

platelets, endothelial cells, leukocytes, clotting factors and the fibrino-lytic process.

- Markers of cellular and immune activation are needed. Measurements of soluble adhesion molecules in plasma and, possibly, of cleavage products in urine, might improve the diagnosis of CVD and the evaluation of its prognosis. In addition these markers may help to establish and monitor the impact of nutrient supplementation.

4.4.4.4 Human intervention studies

- Interactions and relative amounts of fatty acids (*e.g.* the n − 3 PUFA:n − 6 PUFA ratio, palmitate) and the effect of dietary cholesterol on lipid metabolism (particularly on postprandial hyperlipidaemia), should be studied in well-controlled dietary trials. These investigations should examine systemic trafficking of the fatty acids and the lipoproteins. Effects on other lipid parameters, such as the key enzyme activities in TAG metabolism and lipoprotein particle sizes, should also be studied in order to understand the dietary effects on lipoprotein metabolism.
- The effect of selected dietary components (*e.g.* individual n − 3 and n − 6 fatty acids and combinations of them, antioxidants, fibre) on the processes involved in arterial thrombus formation should also be established and characterized.
- The effect of dietary constituents, *e.g.* fatty acids and bioactive amines, on nitric oxide (NO) metabolism, free radicals and other local endothelial mediators of vascular tone should be explored.
- Particular attention ought to be paid to study design and to characteriza-tion of experimental subjects. Genetic polymorphisms and diet–gene interactions, life-style effects and endocrine factors must all be considered.

4.4.4.5 Evidence of safety

- The potential toxicity of very high intakes of long-chain PUFA and antioxidants should be investigated.

4.5 Intestinal Physiology (see reference 1, pages S147–S170)

4.5.1 Introduction. The gut is an obvious target for the development of functional foods since it acts as an interface between the diet and the events that sustain life. The target functions relating to the physiology of the gut can be subdivided according to those that determine transit time and stool characteristics, those that involve the colonic microflora, those that are associated with the gut-associated lymphoid tissue (GALT), and those that depend on the products of nutrient fermentation.

The development of intestinal and mucosal microflora provides the basis for the gut barrier that prevents pathogenic bacteria from invading the gastrointestinal tract. The balance of intestinal microflora together with the gut immune system allows the resident bacteria to have a protective function.

The principal role of the intestinal flora, apart from their barrier function against infection, is to salvage energy from carbohydrates not digested in the upper gut, through fermentation. The main substrates for fermentation are endogenous (*e.g.* mucus) and dietary carbohydrates that have escaped digestion in the upper gastrointestinal tract. These include starch that enters the colon (resistant starch), as well as non-starch polysaccharides, *e.g.* celluloses, hemicelluloses, pectins and gums. Other carbohydrate sources available for fermentation include non-digestible oligosaccharides, and sugar alcohols. In addition, proteins and amino acids can be effective as growth substrates for colonic bacteria. Total substrate availability in the human adult colon is 20–60 g of carbohydrate and 5–20 g of protein per day.

4.5.2 Key target functions

4.5.2.1 Optimal intestinal function and stool formation. Both large bowel integrity and the colonic microflora are important in determining the characteristics of the stool. Such characteristics include stool weight and consistency, stool frequency and total intestinal transit time. These are possibly the most reliable markers of general colonic function.

4.5.2.2 Colonic flora composition. Bacterial numbers and composition vary considerably along the human gastrointestinal tract but the large intestine is by far the most intensively populated microbial ecosystem with several hundred species accounting for a total of between 10^{11}–10^{12} bacteria per gram of contents. Quantitatively, the most important genera of intestinal bacteria in man are the bacteroides and the bifidobacteria, which can account for 35 % and 25 % of the known species, respectively. The microflora of the large intestine is acquired at birth and reflects the breast-milk-based diet of infants. Subsequently, however, differences in the species composition develop, largely as a result of differences in adult-type diets.

The colonic microflora is a complex interactive community of organisms and its functions are a consequence of the combined activities of the microbial components. The group of potentially health-promoting bacteria are thought to include principally the bifidobacteria and lactobacilli. The major diseases in which changes in the composition of the gastrointestinal microflora might possibly play a role are gastrointestinal infections, constipation, irritable bowel syndrome, inflammatory bowel diseases and colorectal cancer.

The composition and activities of the colonic microflora have to be considered both as a risk indicator and a risk factor for large bowel functions (see Section 3). The composition and activity of the faecal flora, when analysed correctly, are also good indicators of the residual colonic flora. Traditional gut microbiological methodologies are based on morphological and biochemical properties of the organisms. However, recent advances in molecular genetics for quantitative and qualitative monitoring of the nucleic acids of human colonic microflora have revolutionized their characterization and their identification. This approach also allows extensive (including multiple centre) bacteriological investigations.

4.5.2.3 Control of gut-associated lymphoid tissue (GALT) function. The human intestine GALT represents the largest mass of lymphoid tissue in the body, and about 60 % of the total immunoglobulin produced daily is secreted into the gastrointestinal tract. The colonic microflora is the major antigenic stimulus for specific immune responses at local and systemic levels. Abnormal intestinal response to a foreign antigen, as well as local immunoinflammatory reactions, might, as a secondary event, induce impairment of the intestine's function because of breakdown of the intestinal barrier.

4.5.2.4 Control of fermentation products. The importance of fermentation products in the form of SCFA (namely butyrate, acetate and propionate) for colonic health has been increasingly documented. Butyrate is the most interesting of the SCFA since, in addition to its trophic effect on the mucosa, it is an important energy source for the colonic epithelium and regulates cell growth and differentiation.

4.5.3 Examples of options for modulation

- Probiotics, prebiotics and possibly synbiotics are interesting concepts to support the development of functional foods that are targeted towards gut functions. A probiotic is defined as 'A live microbial food ingredient that is beneficial to health'. Prebiotics, in contrast, are 'non-digestible food components that beneficially affect the host by selectively stimulating the growth and/or activity of one or a limited number of bacteria in the colon, that have the potential to improve host health'. Synbiotics are best described as 'a mixture of probiotics and prebiotics that beneficially affects the host by improving the survival and implantation of live microbial dietary supplements in the gastrointestinal tract'.
- Probiotics have been shown to stimulate the activity of some compounds of the GALT, *e.g.* immunoglobulin A (IgA) antibody response, production of cytokines and reduction of the risk of rotavirus infections.
- Prebiotics (as well as probiotics) help the colonic microflora to reach/ maintain a composition in which bifidobacteria and/or lactobacilli become predominant in number. This is considered optimal for health promotion.
- Non-digestible carbohydrates increase faecal mass both directly by increasing non-fermented material and/or indirectly by increasing bacterial biomass; they also improve stool consistency and stool frequency.
- Promising food components for functional foods will be those, such as specific non-digestible carbohydrates, that can provide optimal amounts and proportions of fermentation products at relevant sites in the colon, particularly the distal colon, if they are to help reduce the risk of colon cancer.

4.5.4 Research opportunities

4.5.4.1 Basic science

- More methods need to be developed to characterize intestinal microflora, including the non-culturable species. Molecular approaches based on the separation and identification of genetic material, such as 16S ribosomal RNA, are promising.

Table 5 Examples of opportunities for modulation of target functions related to intestinal physiology by candidate food components with possible markers*

Target functions	Possible markers	Candidate food components
Optimal intestinal functions and stool formation	stool consistency stool weight stool frequency transit time	non-digestible carbohydrates probiotics prebiotics synbiotics
Colonic flora composition	composition enzyme/metabolic activities	probiotics prebiotics synbiotics
Control of GALT function	IgA secretion cytokines	probiotics prebiotics synbiotics
Control of fermentation products	short-chain fatty acids	prebiotics synbiotics

* The information given in this table is derived from the Theme Paper (see Bellisle *et al.* 1998, S147–S171), where, if not given in this Consensus Document, further details of options for modulations, markets and safety issues can be found.
GALT = gut-associated lymphoid tissue.
IgA = immunoglobin A.

- We need to understand the role of the human diet in the general modulation of immune functions, particularly the exact role of GALT.

4.5.4.2 Cross-sectional and prospective epidemiological evidence

- Further work is needed to characterize the composition and activities of human colonic microflora to investigate the effects of age and race, and to explore the influence of different dietary habits. Of particular interest would be large-scale studies of different European populations, using molecular approaches.

4.5.4.3 Markers

- Markers related to immune functions associated specifically with GALT need to be developed and validated.
- Improved characterization and validation are needed of the composition of bacterial genera and strains and the activities of the colonic microflora (*e.g.* specific enzymes, carcinogen formation). Only then can these be regarded as markers of important target functions.
- Early markers of carcinogenesis (*e.g.* recurrence of large adenomas or gut mucosal changes such as DNA damage, DNA repair and apoptosis) need to be validated as a prerequisite to investigating the effect of functional foods on the risk of colon cancer.

4.5.4.4 Human intervention studies

- What factors determine the intra-individual variation in response to probiotics and prebiotics?
- Well-designed placebo controlled human studies are required to examine the effects of different probiotic and prebiotic components.

4.5.4.5 Evidence of safety

- The long-term effects of permanent changes in the composition of the colonic microflora need to be monitored.

4.6 Behavioural and Psychological Functions (see reference 1, S173–S193)

4.6.1 Introduction. Functional foods might lead to an improved state of health and well-being and/or reduction of risk of disease. However, there are some foods or food components that are not directly related to disease or to health in the traditional sense but, nevertheless, provide an important function in terms of changing mood or mental state. These foods, therefore, are involved in creating more a sense of 'feeling well' than of 'being well'. Effects on behaviour, on emotional state, and on cognitive performance fall within this category.

Behaviour is probably the most varied and complex of all human responses. The complexity and variability arises from the fact that behaviour is the cumulative outcome of two distinct influences – biological factors (encompassing genetics, gender, age, body mass, etc.) and socio-cultural aspects (including tradition, education, religion, economic status).

4.6.2 Key target functions. The effects of foods on behavioural and psychological functions are conspicuously varied, typically subtle and frequently interdependent. For the purpose of simplicity, however, they can be grouped into four areas of target functions.

4.6.2.1 Appetite, satiation and satiety. Food intake can be affected by sensory, gastrointestinal and metabolic factors, as well as by social cues, all of which influence appetite, either in terms of meal size (satiation) or eating frequency (satiety) or in changes in food preferences and dietary selection.

The most widely used markers for assessing parameters of appetite and satiety are visual analogue scales. These are subjective measurements of sensations such as hunger, desire to eat and fullness.

Objective assessment of energy and/or nutrient intake of individuals is made either directly, by duplicate meal analyses, or indirectly, with diaries of dietary consumption.

Modulation of food intake is most often considered in terms of reducing food intake (to normalize body weight of overweight and obese individuals), but it is also important to consider it in terms of increasing food intake to increase body weight in persons at risk of malnutrition (such as elderly people, people suffering from eating disorders or those recovering from illness). See Section 4 for consideration of markers for body weight control.

4.6.2.2 Cognitive performance. There are several performance markers that are used to assess cognitive function. These range from reactions to single stimulus tests to those involving complex interactive inputs. They can be based on recognition of either numbers (to test vigilance, rapid information processing and/or continuous performance) or words (to monitor immediate recall, word recognition and/or memory span). They can also be based on visual or auditory stimuli (to examine 'simple' or 'choice' reaction times) and on reactions to dynamic stimuli (to analyse psycho-motor performance).

4.6.2.3 Mood and vitality. Target functions relating to mood and vitality have focused generally on behaviours such as sleep and activity (including hyperactivity), as well as feelings of tension, calmness, drowsiness and alertness. These psychological or behavioural states can be assessed either subjectively (with questionnaires and visual analogue scales) or objectively (with electrophysiological, heart rate or blood pressure monitoring systems). Because of the expense and sophistication of electrophysiological instruments, the data gathered on these markers are inevitably restricted to laboratory investigations. Portable continuous monitoring systems for heart rate and blood pressure are ideal for field work.

Changes in mood and vitality as a result of shifts in circadian rhythm (jet lag) are also an important target function.

4.6.2.4 Stress (distress) management. Certain foods or individual nutrients have been implicated in causing or alleviating behavioural stress (or distress). The most appropriate markers for stress states are heart rate, blood pressure, blood catecholamines and skin electrical impedance. With regard to pain, data on blood opioid levels can provide additional useful information.

4.6.3 Examples of options for modulation

- Fat, protein and carbohydrates have independent and differing effects on appetite and satiety; protein has been shown to be the most effective, and fat the least effective, at suppressing energy intake. Macronutrient substitutes (*e.g.* fat substitutes and high-intensity sweeteners), because they improve palatability of diets, might, in certain circumstances, also help to control body weight through a reduction in total energy intake.

- There are general beneficial effects of glucose on various aspects of mental performance, including an improvement in working memory, improved decision time, faster information processing and better word recall. Caffeine can also lead to an improvement in most measures of cognitive performance (reaction time, vigilance, memory and psychomotor performance), particularly during the morning.

- High carbohydrate meals help to produce feelings of drowsiness, sleepiness and calmness. In addition, the amino acid tryptophan reduces sleep latency and promotes feelings of drowsiness and fatigue in both adults and children. Tyrosine and tryptophan might help with jet lag but there is only scant scientific evidence for this effect.

- Alcohol, whose intake is both traditional and widespread in Europe, is one of the very few substances to affect all major areas of psychological and behavioural function (appetite, cognitive performance, mood and stress) and the effects are conspicuously dependent on the dose.

- Sweet foods, such as sucrose, might relieve distress in young infants, as might activation of endogenous opioids (β-endorphins) reduce pain perception in the general population.

Table 6 Examples of opportunities for modulation of target functions related to behaviour and physiological functions by candidate food components with possible markers*

Target functions	Possible markers	Candidate food components
Appetite, satiety and satiation	reliable direct measure of food intake indirect measures of food intake subjective ratings of appetite and hunger	protein fat replacers fat substitutes sugar substitutes structured fats specific fatty acids
Cognitive performance	objective performance tests standardized mental function tests interactive games multifunction tests	glucose caffeine **B** vitamins choline
Mood and vitality	quality and length of sleep attitudes (questionnaires) standardized mental function tests ratings of subjective states	alcohol carbohydrates carbohydrate: protein ratio tyrosine and tryptophan
Stress (distress) management	blood pressure blood catecholamines blood opioids skin electrical impedance heart rate continuous monitoring	alcohol carbohydrates sucrose

* The information given in this table is derived from the Theme Paper (see Bellisle *et al.* 1998, S173–S193), where, if not given in this Consensus Document, further details of options for modulations, markets and safety issues can be found.

4.6.4 Research opportunities

4.6.4.1 Basic science

- It is crucial to improve the reliability of food intake data. The current methods of measurement, both indirect and direct, are either inaccurate (both under- and over-estimations), tedious or expensive. A more accurate and precise method is required, which is not only relatively inexpensive and user/operator friendly but also can be used for large field trials. The use of digital cameras, and the associated computer technology needed for image analysis, might make it possible to measure food intake more reliably under realistic free-living conditions. Patterns of dietary intakes of food components throughout Europe can then be obtained to reflect the cultural diversity.
- Is the ability of alcohol to correct depressive episodes linked, qualitatively and quantitatively, to central neurochemical activity, such as that of serotonin?
- 4.6.4.2 Cross-sectional and prospective epidemiological evidence
- The large population of Europe, and the ethnic and cultural diversity within it, offer important opportunities to test the validity of laboratory investigations and to study regional differences in psychotropic effects of foods.

4.6.4.3 Markers

- Better measures of cognitive function (memory, reaction time, alertness) are needed. Testing of multiple, rather than independent, responses to nutritional stimuli is necessary to understand the interaction between cognitive processes. This requires development, elaboration and evaluation of methods that can assess operational performance in complex conditions. Driving or flying simulators, and interactive electronic games, are examples of such methods.
- It cannot be assumed that identical foods will exert similar effects in every individual. Methods are needed to identify vulnerable or sensitive individuals. Is it possible to develop genetic markers?

4.6.4.4 Human intervention studies

- Behavioural animal studies are no substitute for human investigations. Data derived from animal studies give, at best, only an insight into the possibility that similar phenomena occur in man. Even when a potential effect has been demonstrated in man in the laboratory, it is very important to proceed with appropriate human trials in a free-living situation.
- Most research into behaviour and psychological functions is short term. Extended observations, beyond the immediate effect, are required to see if there is long-term adaptation to the intake of specific food components. In studies on satiety, for example, there are indications that short-term effects of food might weaken with repeated administration.

4.6.4.5 Evidence of safety

- Evidence of safety of food components is particularly important in the study of behaviour since this encompasses intrinsic and extrinsic aspects –

the classical biochemical and histopathological changes related to ingestion of specific food components and the changes arising from modification of the person's behaviour or mood. The safety issue of alcohol is a
good example of these two aspects.

- In some cases the influence of dietary substances on behaviour and
psychological functions involves either doses above those normally consumed in food, or doses causing some degree of nutritional imbalance.
The effect of melatonin on circadian rhythms, or of L-tryptophan on
sleep, fall within this area.
- Long-term consequences are as important as those observed in the short
or medium term. For example, the modulation of eating behaviour
through the modification of appetite or satiety should never lead to a
deterioration in the general nutrition status of individuals.

5 Technology Aspects of Functional Food Science[2]

5.1 Introduction

The traditional aim of food processing is to convert raw materials into edible,
safe, wholesome, nutritious food products with desirable physio-chemical
properties, extended shelf life and optimal features for palatability and
convenience of eating. The development of functional foods, however, requires
an additional aim – namely, that the functional component (as defined in this
document) is either created or optimized (see below). This will require an
increased level of complexity and monitoring of food processing.

New raw materials will have to be considered, including those produced by
genetic modification and emerging thermal and non-thermal technologies.
Safety issues then become increasingly important (*e.g.* inhibiting transfer of
secondary genetic material in genetic modification and effective inactivation of
specific components such as toxins, anti-nutrients, micro-organisms, allergens
or enzymes by gentle processing techniques).

Integration throughout the entire food chain will also be required to ensure
product safety and quality, as well as preservation and/or enhancement of
functional components. Sharing of knowledge with scientists in related areas,
such as nutrition, is essential to change the traditional pattern of thinking in
food science and technology.

The three key areas for technological challenges that have been identified
are:

1. The creation of new functional food components in traditional and new
raw materials and by *de novo* synthesis.

2. The optimization of functional food components in raw materials and in
foods (*e.g.* maximal preservation or retention of components, modification of
their function and their increased bioavailability).

3. The effective monitoring of the amount and efficacy of functional food
components in raw materials and in foods.

Examples will be drawn from the five technology areas covered in the Technology Theme Group papers (*i.e.* antioxidants, minerals, micro-organisms, carbohydrates and proteins). Some examples of research opportunities (posed as questions in italics) to meet the technological challenges will also be selected from these areas.

5.2 Examples of Technological Challenges and Research Opportunities for the Creation of New Functional Food Components

● To identify viable technologies for the use of new raw materials as sources.

Can genetic modification be used to increase the antioxidant content of plant foods?

Can under-utilized and unconventional sources, such as algae and seaweeds, be used as raw materials for the extraction of carbohydrates?

● To produce new classes of functional food components, which can be incorporated into foods and used as substrates for beneficial micro-organisms.

Can we find a wider spectrum of sources for production of non-digestible oligosaccharides or other carbohydrates? Can we use 'tailored' degradation and transglycolysation of suitable oligosaccharides?

● To generate new functional carbohydrates (*e.g.* resistant starches and non-digestible carbohydrates) by upgrading raw materials and by-products, which are rich in carbohydrates, using extraction, fractionation and chemical or physical treatments.

Can the cell wall matrix of the raw plant materials be modified to alter the binding of water, cations and bile salts to polysaccharides? Can new fat replacers be produced in this way?

● To develop the large-scale isolation and enrichment of bioactive peptides and proteins from various raw materials, such as milk and plants, using enzymatic and bacterial hydrolysis using separation and chromatographic techniques.

Can bioreactors based on immobilized proteolytic enzymes or live micro-organisms be developed?

5.3 Examples of Technological Challenges and Research Opportunities to Optimize the Amount and Efficacy of Functional Food Components

● To develop economic membrane-processing techniques and to overcome problems of membrane fouling and efficiency to obtain maximum recovery of functional food components,

Is it possible to develop membrane-processing techniques, such as membrane

purification for fats and oils to reduce energy consumption, minimize degradation and to maximize stability and antioxidant potential?

Can we use membrane technologies on food process wastes to recover absorbable mineral complexes with food grade ligands?

- To understand the relationships between packaging environments and antioxidant retention to maximize the functionality of antioxidants in processed foods.

Can we obtain better data on the antioxidant content of raw materials and of minimally processed foods to ensure their maximal retention?

What is the impact of controlled and modified atmosphere packaging environments on antioxidant retention?

- To use separation technology to provide specific fractions from raw materials, each with predetermined mineral content and bioavailability.

Can the controlled separation of skimmed milk from butter preferentially retain important minerals in the skimmed milk?

- To ensure that new minimal processing technologies not only improve the stability of the product, but also retain or improve, mineral availability and allow the release of 'absorbable' minerals.

Can mineral content, yield, speciation, solubility and bioavailability be improved using gentle physical processes, such as high hydrostatic pressure treatment, ultrasound treatment, high-intensity electric field pulse technology, membrane techniques, fractionation technologies (such as liquid–liquid phase separation for partitioned foods) and enzymic processes (such as malting, fermentation and enzyme addition)?

- To modify fermentation processes for micro-organisms to retain their viability during food processing, storage or preparation and thus provide optimal probiotic function.

Can non-thermal technologies (e.g. high hydrostatic pressure, supercritical carbon dioxide treatment or high-intensity electric field pulses) be developed to achieve adequate safety, quality, retention and functionality of products containing probiotics?

- To develop probiotics with increased resistance to the environment within the human intestinal tract.

Is it possible to control fermentation rate in the colon by selecting and developing suitable non-digestible oligosaccharides? Is the fermentability rate due to their degree of polymerization, the type of sugar and glycosidic linkage, the degree of branching or other factors?

Can genetic modification be used to develop highly selective and specific probiotics in terms of their binding properties to intestinal receptors? Can strain selection, fermentation techniques or 'protection' techniques (e.g. by acid-resistant, encapsulated immobilization systems) be used to retain high initial microbial viability and productivity in the intestine?

Can emerging technologies, such as pressure-assisted freezing in conjunction with antifreeze proteins and cryo-protectants, be used as alternatives to conventional freeze-drying of microbial starter cultures for increased viability?

- To optimize the beneficial effects of new carbohydrates (*e.g.* non-

Table 7 Examples of technological challenges, with possible solutions and examples of applications, to optimize functional food components*

Technological challenges	Possible technological solutions	Examples of applications
Creation of functional components from raw materials and from *de novo* synthesis	immobilized enzyme systems membrane processes	bioactive peptides antioxidants minerals
Optimization of functional food components by increasing their concentrations in raw materials	fermentation, enzyme technologies non-thermal processes (*e.g.* high pressure)	minerals antioxidants
Optimization of functional food components through their modification	tailored enzymatic processes	oligosaccharides (fat replacers)
Optimization of functional food components through increased bioavailability	fermentation technologies membrane permeabilization processes (*e.g.* enzymes, electric field pulses)	micro-organisms minerals
Optimization of functional components in raw materials and in foods through maximal retention	encapsulation processes sphere packaging technologies	micro-organisms bioactive peptides antioxidants minerals
Monitoring the production of functional foods and functional components	sensors/markers	micro-organisms minerals carbohydrates

* The information given in this table is derived from the Technological Theme papers (see Knorr *et al.* 1998), where, if not given in this Consensus Document, further details of technological challenges, possible technological solutions and examples of applications can be found.

digestible carbohydrates, fat replacers, resistant starch, soluble/insoluble fibre).

How can the glycaemic index and the resistant starch content be optimized by the careful selection of raw materials and processing conditions?

- To optimize the processing conditions for retaining the activity of bioactive proteins and peptides so that interactions with other food components do not affect the structural/textural quality of food products during processing and subsequent storage.

What are the effects of conventional versus emerging processing technologies on the bioactivity/reactivity of proteins and peptides? Can different delivery systems such as liposomes, microencapsulation systems and emulsion systems optimize their physiological functions?

- To develop alternative production methods for bioactive peptides and proteins from raw materials to avoid problems such as non-availability, which can occur with conventional processing technology.

What are the functional properties and interactions of bioactive proteins and peptides in different food systems, e.g. liquid or solid products (i.e. products with high or low water activity)?

5.4 Examples of Technological Challenges and Research Opportunities to Monitor the Effective Production of Functional Food Components

- To develop efficient technologies to monitor specific functional effects of foods and food components throughout the food chain.

Is it possible to monitor microbial viability and productivity for optimal probiotic function?

- To improve analytical assays to monitor the availability of functional food components at all stages of processing.

Can encapsulated minerals and vitamins retain stability in foods during processing as well as being available for absorption in the intestinal tract?

Can highly specific and sensitive markers be developed that record speciation changes and interactions with food components during processing?

- To develop methods to monitor controlled release and to control bioconversion processes.

Is it possible to monitor the extent of conversion of individual oligosaccharides and the relevant aspects of the fermentation process?

6 Communication of the Health Benefits of Functional Foods

6.1 Introduction

As the relationship between nutrition and health gains public acceptance and as the market for functional foods grows, the question of how to communicate the specific advantages of such foods becomes increasingly important. Communication of health benefits to the public, through intermediaries such as health professionals, educators, the media and the food industry, is an essential element in improving public health and in the development of functional foods. Its importance also lies in avoiding problems associated with consumer confusion about health messages. Of all the different forms of communication, those concerning the use of 'claims' – made either directly as a statement on the label or package of food products, or indirectly through secondary supporting information – remain an area of extensive discussion. It is therefore essential that claims, which are an inherent part of functional foods, must be based on, and driven by, scientific information.

6.2 General Principles of Claims

The fundamental principle that any claim must be true and not misleading should apply equally to those related to health benefits. All such claims, therefore, should be scientifically valid, unambiguous, and be clear to the consumer. The key issue, however, is how this basic principle should be safeguarded without becoming a disincentive either for the industrial development and production of functional foods (an important determinant in trying to achieve the goal of improved public health) or the acceptance of these foods by consumers (the ultimate target for the functional benefit).

6.3 Current Definitions of Claims

One of the difficulties in the communication of the benefits of functional foods is that the term 'health claim' is defined differently in different countries. The meaning of the word 'claim' itself (as opposed to 'health claim') is, however, generally well understood. A widely accepted definition of a 'claim' is that of Codex Alimentarius.[5] It is defined as 'Any representation, which states, suggests or implies that a food has certain characteristics relating to its origin, nutritional properties, nature, production, processing, composition, or any other quality'.

With the term 'health claim', however, there are some appreciable differences in interpretation. The Food Advisory Committee (of the UK Ministry of Agriculture, Fisheries and Food – MAFF), for example, has defined health claim as 'any statement, suggestion or implication in food labelling and advertising (including brand names and pictures) that a food is in some way beneficial to health, and lying in the spectrum between, but not including, nutrient claims and medicinal claims'. In contrast, in the USA, a health claim refers to any statement 'that expressly or by implication characterizes the relationship of any substance to a disease or health-related condition'. In the UK, therefore, a claim related to a disease would be considered as a medicinal claim whereas in the USA it would be regarded as a 'health claim'.

6.4 Current Classification of Types of Claims

The principal difficulty (implied in the above UK definition) in dealing with statements about health benefits is that there are a number of distinct types of claims concerning the association between nutrition and health, and that the definition of them is not the same in all countries.

Codex Alimentarius[6] has very recently classified and defined some of the different forms of claims. For ease of reference we have called these Types 1–4 and they are:

Type 1. Claims related to dietary guidelines or healthy diets
These relate to the pattern of eating in dietary guidelines officially recognized by the appropriate national authority.
>   Examples: Diets low in saturated fats are recommended by ...; The advice of ... is to choose a diet high in fibre.

Type 2. Nutrient content claims (subsection of nutrition claims)
These are nutrition claims that refer to the level of a nutrient contained in a food.
>   Examples: Source of calcium; High in fibre; Low in fat.

Type 3. Comparative claims (subsection of nutrition claims)
These make a comparison of the nutrient level of two or more foods.
>   Examples: Reduced; Less than; Fewer; Increased; More than.

Type 4. Nutrient function claims (subsection of nutrition claims)
This is a form of claim that refers to the physiological role of a nutrient in its relationship to growth, development and normal functions of the body. Claims in this category are similar to 'structure/function' claims in the USA. They make no reference whatsoever to a specific disease, pathological state or abnormal condition.
>   Examples: Calcium might help the development of strong bones and teeth; Contains folic acid; folic acid contributes to the normal development of the fetus.

The above four categories of claims refer to known nutrients and their role in growth development and normal functions of the body. They are, therefore, factual in that they are based on established knowledge that is widely accepted within the scientific nutrition community and make no reference to a particular consequence over and above that expected from the consumption of a balanced diet. Although there is need for research to continue in the areas covered by these claims (to enlarge on the detail of the processes) there is little further necessity to substantiate the concepts themselves.

6.5 Claims Relevant to Functional Foods

This EU Concerted Action supports the development of two further types of claims, which are not covered by the above Codex classification and which differ significantly from them. They embrace, but are not restricted to, claims for functional foods. They are based on the scientific classification of markers for target functions that have been developed in this Consensus Document (see Figure 1 in Section 3). Claims must always be valid in the context of the whole diet and must relate to the amounts of foods normally consumed. This applies to specific effects of foods and food components, both nutrients and non-nutrients.

Type A. 'Enhanced function' claims
These claims concern specific beneficial effects of nutrients and non-nutrients on physiological, psychological functions or biological activities beyond their established role in growth, development and other normal functions of the body.

This type of claim is also similar to a 'structure/function' claim in the USA and, again, makes no reference to a particular disease or pathological state. However, reference to a mild abnormal condition, as for example indigestion or insomnia, could possibly be permitted.

Examples: Certain non-digestible oligosaccharides improve the growth of a specific bacterial flora in the gut. Caffeine can improve cognitive performance.
Folate can help reduce plasma homocysteine levels.

Type B. 'Reduction of disease-risk' claims
Claims for reduction of disease risk relate to the consumption of a food or food component that might help reduce the risk of a specific disease or condition because of specific nutrients or non-nutrients contained within it. These claims correspond to those referred to as 'health claims' in the USA.

Examples: Folate can reduce a women's risk of having a child with neural tube defects. Sufficient calcium intake may help to reduce the risk of osteoporosis in later life.

It is frequently suggested that there should be an additional and specific claim for an improved state of health and well-being. However, since an improved state of health and well-being results from either an enhanced function (physiological or psychological) or from the reduction of risk of disease, it is sufficiently covered by the above two categories.

Satisfactory control procedures are required to ensure that claims are used to their best effect, both for the promotion of public health and protection of the consumer against false and misleading information. Currently Type 1, 2 and 3 claims are generally allowed in most countries. To a much lesser extent, this is also true for Type 4 and Type A claims whilst, in contrast, Type B claims are currently almost universally disallowed (see Section 2).

6.6 Scientific Basis for Claims Relevant to Functional Foods

Perhaps the most pertinent aspect in communication of health benefits is that any claim, or statement, must be based on sound scientific evidence that is both objective and appropriate. However, it is not always clear what constitutes objective and appropriate evidence. One of the major issues to be resolved, therefore, concerns the biological level at which evidence can be accepted as demonstrating functional efficacy.

Data, obtained at several levels of biological organization (molecular, sub-cellular, cellular, tissue, organ, whole body and population level) can be grouped into three types of experimental evidence – biological (biochemical) observations, epidemiological data and intervention trials. All three types of evidence are based on markers. For any given specific food or food component, supporting evidence for functional efficacy might not be available from all three areas. The effectiveness of certain antioxidants in the reduction of risk for cancer, for example, is adequately supported by biological data (at the molecular, sub-cellular and cellular levels), less well supported by epidemiological evidence (at the population level) and hardly at all by clinical intervention trials (whole-body level). In contrast, the effects of dietary fibre are better documented and understood in terms of epidemiology and intervention than they are at the biochemical level.

Whatever judgement or decision is made for establishing proof of functional efficacy, it is important that the required supporting evidence should:

- be consistent in itself;
- be able to meet accepted scientific standards of statistical and biological significance;
- be plausible in terms of the relationship between intervention and results;
- be provided from a number of sources, including human studies.

Clearly, the use of different types of markers can play a very important role in justifying claims. This EU Concerted Action proposes that the classification of markers on the basis of their relationship to the target function or disease endpoint (see Section 3) now offers a scientific approach for the basis of Type A and Type B claims:

If evidence for the effects of a functional food or a functional food component is based on a marker of target function or biological response (functional markers), then a Type A claim (enhanced function claim) might be justified.

A Type B claim (reduced risk of disease claim), however, would only be justified if the evidence for the effect of a functional food or a functional food component is based on an intermediate endpoint marker of disease. This marker would have to be shown to be significantly and consistently modulated by the functional food component for the evidence to be acceptable.

6.7 Linking the Scientific Basis of Functional Foods with the Communication About their Benefits to the Public

The development of functional foods, with their accompanying claims, will proceed hand in hand with progress in food regulation which is the means to guarantee the validity of the claims as well as the safety of the food. Science in itself cannot be regulated and functional food science provides only the

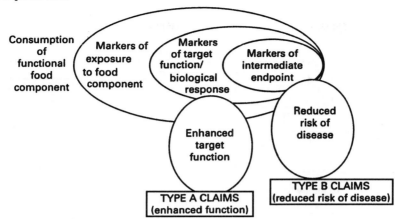

Figure 2 *From scientific evidence based on markers for functional foods to types of claims relevant to them*

scientific basis for these regulations. At the present time, in several European countries, different initiatives have been taken, based on the dialogue between the food industry, retailers, regulatory authorities with consumer organizations. A self-regulating programme for 'health claims', which is not independent of existing legislation or the requirements of relevant authorities but would complement both, has been introduced in Sweden and more recently in The Netherlands. It is also being actively considered in the UK and Belgium.

This EU Concerted Action has proposed a scheme whereby the scientific basis of functional food development can be linked to the communication of the benefits of functional foods to the public. This scheme is summarized in Figure 3. Improved communication with consumers should follow if the principles in this scheme are adopted.

7 Executive Summary

1. The Functional Food Science in Europe (FUFOSE) project was introduced, evaluated and accepted by the EU DG XII FAIR Programme as a Concerted Action. Its aim was to develop and establish a science-based approach for the emerging concepts in functional food development. Over the last three years of this EU Concerted Action co-ordinated by ILSI Europe, scientific data have been evaluated and new concepts have been elaborated. This Consensus Document is the culmination of the EU Concerted Action and its key points and recommendations are summarized here. It is by no means the end of the process, but, rather, an important starting point and the stimulus for functional food development.

2. Considerable progress has been made in scientific knowledge leading to the identification of functional food components which might eventually lead to an improved state of health and well-being and/or reduction of risk of disease. Consumers are becoming more aware of this development as they seek

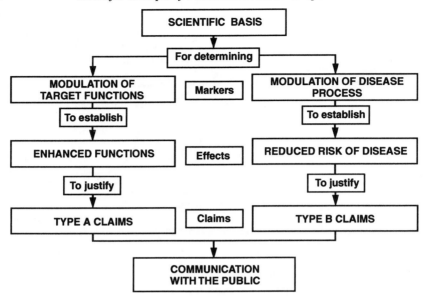

Figure 3 *From the scientific basis for functional foods to communication with the public*

a better-quality, as well as a longer, life. The food industry has an opportunity to provide products that are not only safe and tasty, but also functional. The originality of the approach in this EU Concerted Action is that it is function-based, rather than product-based. The latter approach would have to be influenced by local considerations of different cultural as well as dietary traditions, whereas the function-based approach starts from the biologically based science that is universal. Furthermore, and most importantly, the function-based approach in this EU Concerted Action has allowed the development of ideas that suggest a unique way in which to link this scientific basis of functional foods with the communication about their possible benefits to consumers.

3. This EU Concerted Action has adopted the following working definition, rather than a firm definition, for functional foods:

A food can be regarded as 'functional' if it is satisfactorily demonstrated to affect beneficially one or more target functions in the body, beyond adequate nutritional effects in a way that is relevant to either an improved state of health and well-being and/or reduction of risk of disease.

4. Functional foods must remain foods and they must demonstrate their effects in amounts that can normally be expected to be consumed in the diet. They are not pills or capsules, but part of a normal food pattern. A functional food can be a natural food, a food to which a component has been added, or a food from which a component has been removed by technological or biotechnological means. It can also be a food where the nature of one or more components has been modified, or a food in which the bioavailability of one or more components has been modified; or any combination of these possibilities.

A functional food might be functional for all members of a population or for particular groups of the population, which might be defined, for example, by age or by genetic constitution.

5. The development of functional foods must rely on basic scientific knowledge of target functions in the body that are relevant to an improved state of health and well-being and/or the reduction of risk of diseases, the identification of validated markers for these target functions and the evaluation of sound scientific data from human studies for their possible modulation by foods and food components. This EU Concerted Action has proposed that markers can be classified according to whether they are markers of exposure to the functional food component, whether they are markers that relate to target function or biological response or whether they are intermediate markers of the actual disease endpoint or health outcome (see Figure 1).

6. Consumers must be made aware of the scientific benefits of functional foods and this requires clear and informative communication through messages (claims) on products and in accompanying materials. This EU Concerted Action has identified two types of claims that are vital to functional food development and has provided a scientific basis for them to help those who have to formulate and regulate the claims (see Figure 2).

Claims for 'Enhanced Function Claims' (Type A) should require that evidence for the effects of the functional food is based on establishment and acceptance of validated markers of Improved Target Function or Biological Response, while claims for the Reduced Risk Of A Disease (Type B) should require that evidence is based on the establishment and acceptance of Markers of Intermediate Endpoints of Disease. These markers must be shown to be significantly and consistently modulated by the functional food or the functional food component for either type of claim to be made. This EU Concerted Action has therefore proposed a scheme whereby the scientific basis of functional food development can be linked to the communication of their benefits to the public (Figure 3). If the principles of such a scheme can be universally adopted, then this should ultimately improve communication to consumers and minimize their confusion.

7. Functional foods must be safe according to all standards of assessing food risk and new approaches to safety might need to be established. This EU Concerted Action proposes that the development of validated markers as described above should, if possible, be used and integrated in the safety assessment with particular attention being paid to long-term consequences and interactions between components.

8. The development of functional foods, with their accompanying claims, will proceed hand in hand with progress in food regulation, which is the means to guarantee the validity of the claims as well as the safety of the food. Science in itself cannot be regulated and functional food science provides only the scientific basis for these regulations.

9. The Individual Theme Group papers, which are the science base for this Concerted Action, represent the critical assessment of the literature by European experts.[1,2] The health outcomes that have been covered are those

related to growth, development and differentiation, substrate metabolism, defence against reactive species, cardiovascular system, intestinal physiology and behavioural and psychological functions.

10. This Consensus Document has selected some examples from each theme paper to illustrate the key target functions for each health outcome with their possible markers and candidate functional food components and has also indicated some research opportunities, technological challenges and safety considerations. Among the technological challenges are issues relating to the creation of new functional food components, their optimization within foods and the continual monitoring of their efficacy throughout food processing.

11. The collaboration between the many disciplines involved in food and nutrition science, such as that which has existed throughout this EU Concerted Action, is essential for successful innovation in functional food development leading to an improved state of health and well-being and/or reduction of risk of disease for consumers.

12. The European Union must play a leading role in the development of, and communication about, functional food science and the European Food industry should share the lead and work in partnership with the scientific community. Support must be given to integrated research programmes aimed at solving the key scientific and technological challenges that have been identified in this Consensus Document.

8 Key Messages

- The food industry has unique opportunities to develop products that are not only nutritional in the traditional sense, but which have additional activity that can lead to an improved state of health and well-being and/or reduction in risk of disease (functional foods).
- Foods can be regarded as functional if they can be satisfactorily demonstrated to affect beneficially one or more target functions in the body, beyond adequate nutritional effects in a way that is relevant to either an improved state of health and well-being and/or reduction of risk of disease. Functional foods must remain foods and they must achieve their effects in amounts that could normally be expected to be consumed in a diet. They are not pills.
- A function-based, rather than a product-based, approach has been proposed whereby the scientific basis of functional foods can be linked to the communication of their benefits to the public. The ability to communicate these benefits is essential for the successful development of functional foods and their role in improving public health.
- Scientific understanding of the way in which components affect body processes involved in health and well-being enables the development of markers that could register the impact of the new food products and could also be used in their safety assessment.
- Evidence from human studies based on markers relating to biological response or on intermediate endpoint markers of disease could thus

provide a sound scientific basis for messages and claims about the functional food products. Two types of claims are proposed that would relate directly to these two categories of markers: enhanced function claims, and reduced risk of disease claims.

● Support is needed for integrated research programmes, with interdisciplinary activity, to solve the key scientific and technological challenges and to exploit the scientific concepts in functional food science.

Acknowledgements

We wish especially to thank all of the individual contributors to this project and Consensus Document for devoting their time and efforts within the required timeframe. Their continuous willingness to ensure the success of this Concerted Action has been sincerely appreciated and ILSI Europe, as co-ordinator of this Concerted Action, is extremely grateful to all of them.

References

Theme papers
1   F. Bellisle, A.T. Diplock, G. Hornstra, B. Koletzko, M. Roberfroid, S. Salminen, W.H.M. Saris. *British Journal of Nutrition*, 1998, **80** (Suppl. 1), S1.
2   Knorr *et al. Trends in Food Science and Technology*, 1998, **9**, 293.
3   Institute of European Food Studies (1996) A pan-European Survey of Consumer Attitudes to Food, Nutrition and Health. Dublin, Institute of European Food Studies.
4   D.A. Jonas *et al. Food & Chemical Toxicology*, 1996, **34**, No. 10, 931.
5   Codex Alimentarius (1991) Second Edition. Codex General Guidelines on Claims. (CAC/GL 1–1979, rev 1–1991). Geneva, WHO.
6   Codex Alimentarius (1997) Guideline for Use of Nutrition Claims (CAC/GL 23–1997, Volume 1A, revised). Geneva, WHO.

Other useful references
A   Food Advisory Committee (1991) Report on its review of food labelling and advertising. Report FDAC/REP/10. HMSO, London.
B   ILSI Europe (1998) Addition of Nutrients to Foods: Nutritional and Safety Considerations. In ILSI Europe Report Series. Brussels, ILSI Europe.
C   F.M Clydesdale, S. Ha Chan. *Nutrition Reviews*, 1996, **54**, S1.
D   F.M. Clydesdale. *Nutrition Reviews*, 1997, **55**, 413.

2 Pre- and Probiotics

Probiotic Bacteria and the Human Immune System

L. O'Mahony,[1,3] M. Feeney,[1] C. Dunne,[1,3] S. O'Halloran,[1]
L. Murphy,[1] B. Kiely,[3] G. O'Sullivan,[4] F. Shanahan[2] and
J.K. Collins[1,2]

DEPT. MICROBIOLOGY[1] AND MEDICINE,[2] NATIONAL FOOD
BIOTECHNOLOGY CENTRE,[3] UNIVERSITY COLLEGE CORK &
DEPT. SURGERY,[4] MERCY HOSPITAL, CORK, IRELAND

Summary

The human colon contains 10^{14} bacteria/gram of luminal contents. The interactions between this flora and the host immune system has resulted in a state of controlled inflammation at mucosal surfaces. Pertubances of this delicate balance may result in disease and distress to the host.

Probiotic bacteria have been reported to exert significant immune-modulatory effects. These effects include modulation of humoral responsiveness, viral resistance and proinflammatory activity within the gastrointestinal tract resulting in protection against infection and mucosal damage. These responses may result in alterations to host cell gene induction and protein expression (*e.g.* cytokines, antigen presentation, transcription factor activation) due directly to the physical binding of the probiotic strains or the presence of their secreted products.

Research at University College Cork has identified a probiotic strain, *Lactobacillus salivarius* subsp. *salivarius* UCC118, which survives passage through the human gastrointestinal tract, modulates the enteric flora and adheres to human epithelial cells. The immune perception of this strain was examined in a human feeding trial. No alteration in serum cytokines or soluble receptor levels were observed between the control and test groups over the trial period. Also, there were no differences in the change from baseline concentrations of serum antibodies specific to UCC118 or total serum antibodies. However, mucosal antibody levels specific to UCC118 significantly increased. These results demonstrate that UCC118 is perceived by the human mucosal immune system but does not induce systemic reactivity.

Currently, we are examining the mechanisms underlying the immune perception of probiotic bacterial strains both *in vitro* and *in vivo*, focusing on control as well as stimulation of the immune response. In summary, we have identified probiotic strains that exert specific immunomodulatory effects within the gastrointestinal tract.

1 Introduction

Both immunological and non-immunological defence mechanisms protect the human gastrointestinal tract from colonisation by intestinal bacteria.[1] Innate defence mechanisms include the low pH of the stomach, bile salts, peristalsis, mucin layers and anti-microbial compounds such as lysozyme.[2] Immunological mechanisms include specialised lymphoid aggregates, underlying M cells called Peyer's Patches which are distributed throughout the small intestine and colon.[3] Luminal antigens presented at these sites result in stimulation of appropriate T and B cell subsets with establishment of cytokine networks and secretion of antibodies into the gastrointestinal tract.[4,5] In addition, antigen presentation may occur *via* epithelial cells to intraepithelial lymphocytes and to the underlying lamina propria immune cells.[6] Therefore, the host invests substantially in immunological defence of the gastrointestinal tract. However, as the gastrointestinal mucosa is the largest surface at which the host interacts with the external environment, specific control mechanisms must be in place to regulate immune responsiveness to the 100 tons of food which is handled by the gastrointestinal tract over an average lifetime.[7] Furthermore, the gut is colonised by over 500 species of bacteria numbering 10^{11}–10^{12}/g in the colon. Thus, these control mechanisms must be capable of distinguishing non-pathogenic adherent bacteria from invasive pathogens that would cause significant damage to the host. In fact, the intestinal flora contributes to defense of the host by competing with newly ingested potentially pathogenic micro-organisms. Furthermore, consumption of non-pathogenic, or probiotic, bacteria has resulted in enhancement of immune parameters in healthy volunteers. Examples of these immune modulatory effects can be observed in Table 1.

Table 1 *Immune enhancing effects following oral consumption of probiotic bacteria*

Observed effect	Reference
Increased macrophage phagocytosis	10
Increased Natural Killer (NK) cell activity	11
Increased IFNg serum levels	12
Increased B cell and NK cell numbers	12
Promotion of IgA responses	11, 13–15
Increased Delayed Type Hypersensitivity (DTH) responses	16

2 Immune Education

The enteric flora are important to the development and proper function of the intestinal immune system. In the absence of an enteric flora, the intestinal immune system is underdeveloped, as demonstrated in germ free animal models, and certain functional parameters are diminished, such as macrophage phagocytic ability and immunoglobulin production.[8,9] The importance of the gut flora in stimulating non-damaging immune responses is becoming more evident. The increase in incidence and severity of allergies in the western world has been linked with an increase in hygiene and sanitation, concomitant with a decrease in the number and range of infectious challenges encountered by the host. This lack of immune stimulation may allow the host to react to non-pathogenic, but antigenic, agents resulting in allergy or autoimmunity.

The deliberate consumption of non-pathogenic bacteria, such as probiotic bacteria, may provide a health promoting immune challenge to the host. The promotion of immunoglobulin secretion into the lumen, following consumption of probiotic bacteria,[14] may bind allergens and prevent their recognition by the host. In addition, interaction with certain intraepithelial lymphocyte subsets may suppress immune responses to allergens within the gastrointestinal tract, possibly resulting in tolerance to oral and inhaled antigens.[17] Inadequate priming of T cell subsets may result in an incorrect cytokine balance or may contribute to a failure of the T cell repertoire to recognise epitopes that are cross-reactive between self and non-self.[18] Thus, the deliberate consumption of probiotic bacteria to replace immune stimuli artificially is being extensively researched at University College Cork.

3 Oral Vaccination by Probiotic Bacteria

The majority of pathogenic organisms gain entry *via* mucosal surfaces. Efficient vaccination of these sites protects against invasion by a particular infectious agent. Oral vaccination strategies have concentrated, to date, on the use of attenuated live pathogenic organisms or purified encapsulated antigens.[19,20] Probiotic bacteria, engineered to produce antigens from an infectious agent, *in vivo*, may provide an attractive alternative as these bacteria are considered to be safe for human consumption (GRAS status).

The LABVAC European network have focussed on the use of lactic acid bacteria as oral vaccine delivery vehicles. Murine studies have demonstrated that consumption of probiotic bacteria expressing foreign antigens can elicit protective immune responses. The gene encoding tetanus toxin fragment C (TTFC) was expressed in *Lactococcus lactis* and mice were immunised *via* the oral route. This system was able to induce antibody titres significantly high enough to protect the mice from lethal toxin challenge.[21] In addition to antigen presentation, live bacterial vectors can produce bioactive compounds, such as immunostimulatory cytokines, *in vivo*. *L. lactis* secreting bioactive human IL-2 or IL-6 and TTFC induced 10–15 fold higher serum IgG titres in mice immunised intranasally.[22] However, with this particular bacterial strain, the total IgA level was not increased by co-expression with these cytokines. Other bacterial strains, such as *Streptococcus gordonii*, are also being examined for their usefulness as mucosal vaccines. Recombinant *S. gordonii* colonising the murine oral and vaginal cavities induced both mucosal and systemic antibody responses to antigens expressed by this bacterium.[23,24] Thus, oral immunisation using probiotic bacteria as vectors would not only protect the host from infection, but may replace the immunological stimuli that the pathogen would normally elicit, thus contributing to the immunological education of the host.

4 Intestinal Bacteria and Gastrointestinal Inflammation

Aberrant immune responses to the indigenous microflora have been implicated in certain disease states, such as inflammatory bowel disease.[25] Antigens associated with the normal flora usually lead to immunological tolerance and failure to achieve this tolerance is a major mechanism of mucosal inflammation.[26] Evidence for this breakdown in tolerance includes an increase in antibody levels directed against the gut flora in patients with IBD. In addition, certain mouse models predisposed to inflammatory lesions in the gastrointestinal tract remain disease free when housed in germ-free conditions or when treated with antibiotics.[27, 28] Certain probiotic bacteria are attractive biotherapeutic agents for the treatment of gastrointestinal inflammation owing to their effects on the composition of the gut flora and activity of the immune system. In a recent clinical trial where pouchitis patients consumed a cocktail of probiotic bacteria, only 15% of the test group relapsed compared to a 100% relapse rate in the placebo group.[29] Therefore, while it is unclear whether bacteria initiate the disease process or if bacterial-induced inflammation is a consequence of intestinal immune dysfunction, probiotic treatment of intestinal inflammation needs to be examined in greater detail, with emphasis on well controlled clinical trials.

5 Selection and Evaluation of Probiotic Strains

The objectives of the Probiotic Research Group, based at University College Cork, were to isolate and identify lactic acid bacteria (LAB) that exhibited beneficial probiotic traits such as immunomodulatory activity. Bacteria were screened for their ability to tolerate bile in the absence of deconjugation activity, acid resistance, adherence to host epithelial tissue, and *in vitro* antagonism of micro-organisms such as pathogens or those suspected of promoting inflammation.[30,31] The strategy adopted for the isolation of potentially effective probiotic bacteria was to screen the microbial population adhering to surgically-resected segments of the GIT (the environment in which they may subsequently be re-introduced and required to function). In total, 1500 bacterial strains from resected human terminal ileum were assessed. From this panel of organisms, *Lactobacillus salivarius* subsp. *salivarius* strain UCC118 was selected for further study, based upon the selection criteria outlined above. In mouse feeding trials milk-borne *Lactobacillus salivarius* strain UCC118 proved capable of successfully colonising the murine GIT.[32] A further trial, approved by the UCC Ethics Committee, was successfully conducted in 80 healthy volunteers (Figure 1). This trial demonstrated that yogurt can be used as a vehicle for delivery of the probiotic strain UCC118 to the human gastrointestinal tract with considerable efficacy in influencing gut flora and colonisation.

Diet	Fermented dairy product exclusion			
		Product Consumption		Normal Diet
Week	0	3	6	9
Day	-21	0	21	42
Timepoint		0	1	2
Sample		F, B, S	F, B, S	F, B, S

Figure 1 *Protocol design for delivery and efficacy in UCC118 feeding trial. F = faecal sample; B = blood sample; S = saliva*

Several immunological parameters were included in the human feeding trial. These included analysis of both systemic and mucosal immune responses. Macrophage phagocytic activity did not change over the consumption period, while granulocyte phagocytic activity significantly increased in the yogurt-borne UCC118-fed group. Systemic antibody titres, both total antibody levels and antibodies specific to UCC118, remained constant. Systemic IL-1, IL-1, IL-2 soluble receptor, IL-6, TNF and IFN levels did not alter throughout the trial period indicating that consumption of UCC118 does not induce significant inflammatory responses in healthy individuals. Mucosal IgA antibodies titres, specific to UCC118, significantly increased in volunteers consuming yogurt-borne UCC118 (Figure 2). Total IgA levels remained unchanged. Thus, consumption of UCC118 in a yogurt product for three weeks was perceived by the mucosal immune system but was not perceived systemically.

6 Conclusion

In summary, we have developed criteria for *in vitro* selection of probiotic bacteria that reflect certain *in vivo* effects on their host, such as modulation of the GIT microflora and modulation of the mucosal immune response, resulting in the production of secretory antibodies specific to the consumed strain. *Lactobacillus salivarius* subsp. *salivarius* UCC118 survives passage through the gastrointestinal tract, adheres to human intestinal cell lines and is non-inflammatory. However, the role that this and other probiotic bacteria play in maintaining intestinal health *via* immunological mechanisms is largely

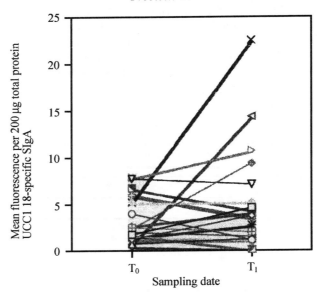

Figure 2 *UCC118LM2-specific sIgA levels in saliva of healthy volunteers before T_0 and after 21 days of test product consumption (T_1)*

unknown. Therefore, both murine and human trials are currently examining the efficacy of probiotic bacteria in the maintenance of gastrointestinal well-being, in both health and disease, with an emphasis on understanding the immunological interactions of these bacterial strains within the gastrointestinal tract.

References

1   V.J. McCracken and H.R. Gaskins, 'Probiotics a critical review', Horizon Scientific Press, UK, 1999, p. 278.
2   D.C. Savage, 'Microbial Ecology of the Gut', Academic Press, London, 1997, p.278.
3   M.F. Kagnoff. *Gastroenterol.* 1993, **105**, 1275.
4   M.R. Neutra and J.-P. Kraehenbuhl, 'Essentials of mucosal immunology', Academic Press, San Diego, 1996, p.29.
5   M.E. Lamm, *Ann. Rev. Microbiol.* 1997, **51**, 311.
6   S. Raychaudhuri and K.L. Rock, *Nat Biotechnol.*, 1998, **16**, 1025.
7   F. Shanahan, 'Physiology of the gastrointestinal tract', Raven Press, 1994, p.643.
8   B.S. Wostmann, 'Germfree and gnotobiotic animal models', CRC Press, Boca Raton, 1996.
9   P.A. Crabbe, H. Bazin, H. Eyssen, and J.F. Heremans, *Int. Arch. Allergy*, 1968, **34**, 362.
10   G. Perdigon, M.E.N. de Macias, S. Alvarez, G. Oliver and A.A. de Ruiz Holgado, *Immunol.*, 1988, **63**, 17.

11 C. de Simone, C. R. Vesley, B. Salvadori, E. Jirillo, *Int. J. Immunother.*, 1993, **9**, 23.

12 C. de Simone, S.B. Bianchi, E. Jirillo, L. Baldinelli, S. di Fabio and E. Jirillo, 'Fermented milks: current research', John Libby Eurotext, London, 1989, p.63.

13 E. Isolauri, H. Majamaa, T. Arvola, I. Rantala, E. Virtanen and H. Arvilommi, *Gastroenterol.*, 1993, **105**, 1643.

14 M. Malin, H. Suomalainen, M. Saxelin and E. Isolauri, *Ann. Nutr. Metab.*, 1996, **40**, 137.

15 H. Link-Amster, F. Rochat, K.Y. Saudan, O. Mignot and J.M. Aeschlimann, *FEMS Immunol. Med. Microbiol.*, 1994, **10**, 55.

16 G. Perdigon and S. Alvarez, 'Probiotics. The Scientific Basis', Chapman and Hall, London, 1992, p.146.

17 Y. Ke, K. Pearce, J.P. Lake, H.K. Ziegler and J.A. Kapp, *J. Immunol.*, 1997, **158**, 3610.

18 G.A. Rook and J.L. Stanford, *Immunol. Today*, 1998, **19**, 113.

19 A.J. Husband, *Vaccine*, 1993, **11**, 107.

20 R.J. Walker, *Vaccine*, 1994, **12**, 387.

21 K. Robinson, L.M. Chamberlain, K.M. Schofield, J.M. Wells and R.W.F. Le Page, *Nat. Biotech.*, 1997, **15**, 653.

22 L. Steidler, K. Robinson, L. Chamberlain, K.M. Schofield, E. Remaut, R.W.F. Le Page and J.M. Wells, *Infect. Immun.*, 1998, **66**, 3183.

23 D. Medaglini, G. Pozzi, T.P. King and V.A. Fischetti, *Proc. Natl. Acad. Sci. USA*, 1995, **92**, 6868.

24 D. Medaglini, M. Oggioni, and G. Pozzi, *Am. J. Reprod. Immunol.*, 1998, **39**, 199.

25 P. Brandzeag, G. Haraldsen and J. Rugtveit, *Springer Semin. Immunopathol.*, 1997, **18**, 555.

26 A. Stallmach, W. Strober, T.T. MacDonald, H. Lochs and M. Zeitz, *Immunol. Today*, 1998, **19**, 438.

27 R. Kuhn, J. Lohler, D. Rennick, K. Rajewsky and W. Miller, *Cell*, 1993, **75**, 263.

28 C. M. Panwala, J.C. Jones and J.L. Viney, *J. Immunol.*, 1998, **161**, 5733.

29 P. Gionchetti, F. Rizzello, A. Venturi, D. Matteuzzi, M. Rossi, S. Peruzzo, G. Poggioli, G. Bazzocchi and M. Campieri, *Gastroenterol.*, 1998, **129**, 1043.

30 J.K. Collins, G. Thornton and G. O'Sullivan, *Int. Dairy J.*, 1998, **8**, 487.

31 C. Dunne, L. O'Mahony, L. Murphy, G. Thornton, D. Morrissey, J.K. Collins, *et al.*, *Am. J. Clin. Nutr.* (in press).

32 J.K. Collins, L. O'Mahony, C. Dunne, D. Morrissey, G.O'Sullivan and F. Shanahan, *Gastroenterol.* (in press).

Communicating the Benefits of Functional Foods to the Consumer

C. Shortt

YAKULT UK, LONDON W3 7XS42, UK

1 Introduction

This is an exciting time for the development of new foods. With advances in food technology and nutritional sciences, food products are reaching the market that are safe, taste good and for which efficacy is supported by sound scientific studies. They not only fulfil nutritional requirements but may have the potential to reduce the risk of disease. Increasingly, consumers are aware that their health and quality of life can be influenced significantly by personal choices, such as diet and exercise. Science provides the foundation for the design of a diet that will promote health and reduce the risk of disease. However, the goal of a healthier future depends on a successful partnership between science, technology and education.[1]

With the opportunity provided by functional foods comes the responsibility to ensure that the message that consumers receive is valid and to ensure that they not only understand the propositions behind functional foods, but also how these foods fit into a balanced diet. A fundamental public health principle is that it is the total dietary balance rather than the consumption of individual products that is critical to positive nutritional health outcomes.

2 The Diet-health Message

If functional foods are to play an effective role in improving human health, several barriers need to be overcome: scientific support for the foods needs to be demonstrated and accepted by a wide audience; consumers must be informed about the benefits of such foods and motivated to buy and incorporate these foods into their diets. Scientific validation of the health benefits prior to launch increases the likelihood of market success and provides a sound basis for communication with the consumer and with medical and health professionals.

The diet–health message is becoming more complex. Translating emerging scientific findings into lay language without loss of accuracy is challenging. There are many reasons why consumers might have difficulty understanding the propositions behind functional foods:

- Consumers have been subjected to a progression of dietary recommendations and are exposed to a plethora of messages that they are often not able to integrate meaningfully.
- They may have pre-existing ideas that prevent them from believing and therefore understanding.
- They may not grasp the use of a concept or term, for example 'phytoestrogens', 'peak bone mass'.
- They may struggle to model or represent mentally the phenomenon or process on which the benefit is based, for example 'cholesterol reduction', 'blood pressure reduction'.

The evolutionary nature of the scientific process can also make it appear contradictory and confusing to consumers, as illustrated poetically by Kritchevsky in *Food News Blues*.[2] In addition, some background information is required if consumers are to be in the position to make informed food choices. The scientific knowledge-base of the population that one is interested in needs to be assessed before any effective communication can be planned; for example in a recent UK study involving 4000 subjects, about 15% were not aware that their digestive systems influenced their general health (Figure 1).

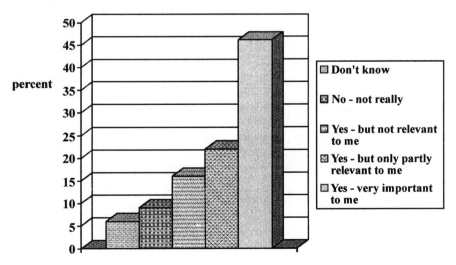

Adapted from (5)

Figure 1 *Response of subjects (n = 4000) to the question: Do you think that the digestive system influences general health?*

3 Resistance to Change

Even when consumers understand a food product proposition, they may not see the relevance of the food to themselves. It has been estimated that there are 137 million 'health uninvolved' consumers in the US alone compared to 53 million 'health aware or active', *i.e.* 137 million people who are resistant to dietary change.[3] These consumers are unlikely to adopt new functional foods even if the benefits of including such foods in the diet have been substantiated and communicated effectively. Similarly, European research suggests that a significant proportion of EU subjects consider that their diet is healthy enough and that they do not need to make changes to the foods they eat, even though health statistics would suggest otherwise.[4] In a recent UK study, over a third of the participants considered that the role of the digestive system in general health was not relevant to them, even though they had a high prevalence of gastrointestinal upsets.[5] Thus, even if a functional food is addressing a relevant 'real physiological need', a proportion of potential consumers will not pick up on the message unless communication strategies are specifically developed to address these issues of relevance and resistance.

4 Communication Channels

There is an ever increasing myriad of channels through which we can convey information about diet and health to consumers. Some of the most frequently cited sources of information on healthy eating include: magazines, TV/radio, newspapers, food packages, health professionals, supermarkets, advertising, books, friends and relatives, industry leaflets (Table 1) and, to a lesser extent, leaflets in clinics, schools and colleges, health food stores, government agencies and consumer organisations. While, as expected, the media represent a major channel of information, it is important to note that other channels also contribute to a significant degree, for example food packages, health professionals, friends and relatives. Indeed, in some European countries the contribution of the media is less dominant than in the UK; for example in France, food

Table 1   *The percentage of subjects in a UK study using sources of information on healthy eating (n = 961) (adapted from reference 6)*

Source	%
Magazines	37
TV/radio	30
Newspapers	28
Food packs	20
Health professionals	19
Supermarkets	18
Advertising	15
Books	15
Relatives/friends	12
Industry leaflets	11

packs, health professionals, and relatives and friends are key sources of information on healthy eating.[6]

Media formats, such as television advertisements, are limited in their ability to convey the often intricate messages that underpin functional foods. Sound-bites, though key to helping engage consumers, are of limited value when it comes to facilitating interpretation of the key health benefits, putting the benefits into context and defining the relevance of the benefits to the individual. However, adverts are useful for introducing concepts and terms, which can be built upon using other communication formats.

While the source of information is established as an important potential influence on credibility, trust/distrust in information may be as important a determinant of consumer reaction as the content of the message. The pattern of used and trusted sources of information on healthy eating is known to differ for each of the EU countries.[6] However, health professionals, government agencies, food packages and TV/radio are generally considered to be trusted sources of such information.

5 Communication in Action

The food industry is a major provider of nutrition information. In a recent survey of key food manufacturers, 96% agreed that consumer education is the most important factor for the success of functional foods.[7] Other factors considered important to success included, in order of importance: taste parity with standard products, proof of efficacy, EC harmonisation on health claims and endorsement by the medical profession. Given the diverse range of food-related information developed and used by food companies, nutritionists and health professionals within the industry can play a decisive role in educating consumers about the health attributes of specific foods in the context of a varied and balanced diet.

In an attempt to raise awareness about probiotic bacteria, Yakult has tackled traditionally taboo consumer issues (bacteria, guts, digestion, wind etc.) and a strongly held perception that all bacteria are bad news.[8] We found that consumers wanted to know more about the digestive system and that there was a lack of available information that explained the workings of the gut in a consumer-accessible format. To address this need, we produced a non-branded lay man's guide to the digestive system *Guide to the Gut*, using the Nutrition Task Force Guidelines on educational materials where appropriate.[9] The *Guide to the Gut* is a great success with consumers and health profes-sionals, and has been described as 'accurate, informative and entertaining'.

6 Conclusion

A multi-faceted approach to communication is essential if the benefits of functional foods are to be understood. While the mass media play a key role in increasing awareness of the benefits, true understanding requires that mass media approaches are complemented with other more community-based channels of communications, such as face-to-face interactions, supermarket demonstrations, workshops and seminars. Such strategies require long-term commitment and investment but, without such a holistic approach to communication with education as the guiding principle, functional foods are unlikely to succeed.

References

1   F. Clydesdale, *Crit. Rev. Fd. Sc.*, 1998, **38** (5), 397.
2   D. Kritchevsky, (1992). Food News Blues –The solution. In: Nutrition, Food and the Environment. Ed. V. Hegarty. Eagan Press. p X111
3   N. Childs, Functional Foods and the Food Industry: developing the incentive to innovate. Presented at the Annual ILSI Meeting, Jan 22–24, 1996.
4   J. Kearney, M. Kearney and M. Gibney, *E. J. Clin. Nutr.*, 1997, **51**, S57.
5   Yakult Omnibus Survey, 1999, Royal Society of Great Britain.
6   M. de Almeida, P. Graca, R. Lappalainen, I. Giachetti, A. Kafatos, A. de Winter and J. Kearney, *E. J. Clin. Nutr.*, 1997, **51**, S16.
7   M. Hilliam, A. James, and J. Young, The European market for functional foods. Leatherhead, 1997.
8   C. Shortt, *Ingred. Health & Nutr.*, 1998, **4**, 30.
9   Guidelines on educational material concerned with nutrition, Dept. of Health, London, 1996.

Functional Foods and the European Consumer

John Young

LEATHERHEAD FOOD RA, RANDALLS ROAD, LEATHERHEAD,
SURREY KT22 7RY, UK

1 Introduction

It is now 5 years since Leatherhead Food RA carried out its ground-breaking
research on consumer attitudes to functional and healthy foods in the UK,
France and Germany. Since then, there have been significant developments in
the market, with the growth of existing categories, such as probiotic and
vitaminised products, and the appearance and marketing of new ingredients,
such as prebiotics and $\omega-3$ oils. Although increasing consumer interest in
health has highlighted an apparent demand for functional foods, and the
number of products on the market has increased, not all by any means have
met with success, despite the use of strong brand names and heavy advertising
support.

As a consequence, it was considered timely to update our knowledge on the
following functional food related issues:

- Consumer perceptions about health;
- Interest and awareness of health claims;
- Functional ingredients and the fortification of foods;
- Functional foods – purchase and attitudes.

Some of the findings from this research, which was based on face-to-face
interviews with a sample of 200 housewives in each of the three major
European markets of the UK, France and Germany, are now given.

2 Consumer perceptions about health

Energy levels were most likely to be used by respondents to judge their own
health in all three countries, with absence of illness taking second place in

France and Germany, and third place in the UK, where physical appearance took second place.

Table 1 also shows that blood pressure and cholesterol levels appear to be more important to German consumers, with 36% mentioning cholesterol levels, compared with 18% in France and just 9% in the UK. Interestingly, awareness of cholesterol levels was much higher on the Continent than in the UK, although it was the French, not the Germans, who were most likely to have had their cholesterol level checked over the previous 12 months. The Germans, however, did have the highest number of respondents (51%) actually being able to say what their cholesterol level was. The UK fared the worst, with just 8% of respondents having had their cholesterol level tested.

Table 1   *Criteria on which respondents judged own health (%)*

	UK (203)	France (199)	Germany (202)
Base			
Energy levels	79	85	81
Physical appearance	67	47	58
Absence of illness	64	66	77
Weight	61	38	57
Bowel movements	30	30	37
Blood pressure	20	29	36
Cholesterol level	9	18	32
Other	2	4	1

Source: Leatherhead Food RA

When questioned on those factors motivating desire to be healthy, it emerged that priorities seemed to vary from country to country (see Table 2). 'To feel good' was the most popular reason given in all three countries. France was notable in that just 7% of respondents cited 'to prevent disease' as a factor motivating desire to be healthy, compared with 59% of UK respondents and 79% of German respondents. In general, the response rates were much higher for UK and German consumers, most of whom selected a number of options. The higher response rate may possibly reflect a greater interest in health and awareness of health issues in these countries.

Table 2   *Factors motivating desire to be healthy(%)*

	UK (203)	France (199)	Germany (202)
Base:			
To feel good	81	79	84
To live longer	70	60	62
To maintain an active lifestyle	60	42	74
To prevent disease	59	7	79
To improve/maintain appearance	56	39	44
To lose/maintain weight	47	29	46
Other	3	–	–

Source: Leatherhead Food RA

When asked to rate the relative importance of three factors, namely diet, exercise and genetic make-up, in their contribution to health, the results differed from Leatherhead's 1993 study, when diet was ranked as the most important factor in all three countries.

In the UK and Germany, respondents still put diet in the lead, it was cited as the most important factor by 62% of respondents in the UK and 59% in Germany, but in France a leading 52% of respondents cited exercise as the most important factor, ahead of genetic make-up, leaving diet in third place (see Table 3).

Table 3 *Factors perceived to be most important to health (%)*

	UK	*France*	*Germany*
Genetic make-up	23	25	17
Exercise	15	52	24
Diet	62	23	59

Source: Leatherhead Food RA

In the UK, the perceived importance of genetic make-up increased between the 1993 and 1998 studies, with 23% of respondents putting it in first place in 1998, compared with just 8% in 1993. The percentage of UK respondents putting exercise in first place also fell, from 24% in 1993 to 15% in 1998, unlike in France and Germany, where the percentage of respondents putting exercise in first place rose from 29% in 1993 to 52% in 1998 in France and from 11% in 1993 to 24% in 1998 in Germany.

As the following table indicates, regarding the main health concerns for self, there are considerable differences from country to country, with breast cancer the only concern in the top five in all three countries, taking first place in the UK and Germany and third place in France.

Table 4 *Top five health concerns for self*

Ranking	*UK*	*France*	*Germany*
1.	Breast cancer	Lack of energy	Breast cancer
2.	Stress	Stress	Lung cancer
3.	Lack of energy	Breast cancer	Bowel/colon cancer
4.	Heart disease	Migraine	Memory decline
5.	Osteoporosis	Coughs/colds/flu	Heart disease

Source: Leatherhead Food RA

When respondents were asked about their main health concerns for their partners, there appeared to be a greater consensus, with heart disease featuring in either first or second place in all three countries (see Table 5).

Response rates to the question of health concerns for children were relatively low, particularly in France, where just over 30% of the sample responded, compared with just over half in Germany and well over half in the UK. Tooth decay and coughs/colds/flu were in the top five concerns in all three countries, while lack of energy, heart disease and bowel disorders were in the top five in two out of the three countries.

Table 5 *Top five health concerns for partner*

Ranking	UK	France	Germany
1.	Heart disease	Stress	Heart disease
2.	Stress	Heart disease	Prostate cancer
3.	Prostate cancer	Prostate cancer	Lung cancer
4.	High blood pressure	Lack of energy	Bowel/colon cancer
5.	Lack of energy	Raised cholesterol	Memory decline

Source: Leatherhead Food RA

Table 6 *Top five health concerns for children*

Ranking	UK	France	Germany
1.	Tooth decay	Coughs/colds/flu	Bowel disorders
2.	Coughs/colds/flu	Lack of energy	Tooth decay
3.	Lack of energy	Bowel disorders	Heart disease
4.	Heart disease	Stress	Coughs/colds/flu
5.	Anaemia	Tooth decay	Bowel/colon cancer

Source: Leatherhead Food RA

Tooth decay was a concern in all three countries, although it was mentioned by only 14% of French respondents, compared with 46% of German respondents and 60% of UK respondents.

When questioned about their ability to influence through diet the medical conditions of most concern, it is interesting to note that in nearly all cases they felt they could exert a high degree of control (see Table 7).

Table 7 *Top five conditions – degree of influence in preventing onset (mean score out of 10)*

UK		France		Germany	
Obesity	8.9	Obesity	9.1	Obesity	9.5
Tooth decay	8.4	Raised cholesterol	8.6	Raised cholesterol	8.9
Raised cholesterol	8.3	Diabetes	8.1	Tooth decay	8.5
Heart disease	7.8	Bowel disorders	8.0	Lack of energy	8.5
Lack of energy/ bowel disorders	7.2	Heart disease	7.7	High blood pressure	7.7

Source: Leatherhead Food RA

Responses in all three countries showed a considerable degree of overlap, with obesity leading in all three countries and raised cholesterol taking either second or third place. Tooth decay, heart disease, lack of energy and bowel disorders were also in the top five in two of the three countries. Interestingly, cancer was the condition where respondents felt they had the least influence through diet.

3 Interest and Awareness of Health Claims

Respondents were shown a list of 25 different health/medical benefits that might be claimed by functional foods and asked which would be of most interest to them. One might expect that interest in claims would be highest for claims in areas where respondents had a high degree of concern, and/or felt they had a considerable influence on prevention.

There was some degree of overlap in the three countries, although 'gives energy' was the only claim in the top five for all three. It is worth noting that in Germany interest in the immune system is very high, cited by 79% of respondents (see Table 8).

Table 8 *Top five health claims (%)*

UK		France		Germany	
Gives energy	73	Gives energy	63	Boosts immune system	79
Promotes healthy bones	71	Lowers cholesterol	53	Promotes healthy bones	70
Promotes healthy teeth	71	Increases resistance to disease	52	Promotes healthy teeth	67
Reduces risk of breast cancer	59	Boosts immune system	50	Gives energy	65
Lowers cholesterol	57	Prevents constipation	49	Promotes a healthy gut	61

Source: Leatherhead Food RA

4 Functional Ingredients and the Fortification of Foods

After a brief explanation of functional foods had been given, respondents were shown a list of ingredients used in them and asked if they were aware of them, if they had bought them, and, finally, which benefits they associated with them.

Table 9 shows the top five ingredients in the UK in terms of products purchased containing them, together with the main benefits associated with each of these ingredients.

When questioned on their attitudes to the fortification of everyday foods with beneficial ingredients, respondents in all three countries were quite positive, particularly in Germany, where a mean score of 4.14 was achieved, with over 80% of respondents agreeing it was a good idea and only 11% disagreeing. In the UK and France, slightly lower mean scores of 3.76 and 3.86, respectively, were achieved. However, these scores still indicated that over 70% of French and UK consumers agreed with the statement that fortification was a good idea. However, there was some concern expressed about the possible danger of overdosing on health-promoting ingredients if they were added to everyday foods.

Table 9   *Top five ingredients in the UK – purchase levels and perceived benefits (%)*

Ingredient	Ever bought	Associated benefits*
1. Polyunsaturated fats	76	Lowers cholesterol
2. Calcium	63	Promotes healthy teeth and bones
3. Iron	60	Prevents anaemia
4. ACE vitamin	57	Promotes healthy bones and teeth/boosts immune system
5. Dietary fibre	55	Prevents constipation

* most popular answers
Source: Leatherhead Food RA

5 Functional foods – purchase and attitudes

From our research it was clear that the term 'functional foods' is still not familiar to consumers, thereby indicating that to the average consumer they are still just 'healthier' alternatives to standard foods.

While health benefit was normally cited as the main reason for purchase, followed by taste, exceptions to this were in fortified confectionery in all three countries, soft drinks in the UK and France, yogurt in the UK and breakfast cereals in Germany, where respondents' liking for the taste scored more highly than health benefits.

When respondents were asked as to who they thought of as credible and trusted suppliers of functional foods, a large number of companies were mentioned.

As can be seen from the following table, Kellogg's was the clear leader in the UK, being mentioned unprompted by 41% or respondents.

Table 10 *Most credible and trusted suppliers of functional foods in the UK (% of respondents mentioned unprompted)*

Kellogg's	41
Heinz	8
Sainsbury	6
St Ivel	6
Tesco	5
Asda	5
Flora	5

Worthy of mention is that there were three retailers in the top six. Unilever will also be gratified to see their Flora brand receiving some credibility. One could conclude that the familiarity of well-known brands/companies does seem to add some reassurance to respondents faced with new products and concepts.

The clear leader in France was Danone (43%) and, in Germany, Nestlé (34%).

In general, respondents were very positive about the idea of functional foods, with more than three-quarters in each country agreeing with a statement that they would buy functional foods for themselves and their families, and over 55% in each country agreeing with a statement that they believed that functional foods would improve their health.

Time constraints have only allowed me to give a brief insight into the findings from our recent research study entitled, Functional Foods and the European Consumer – Implications for Product and Market Development.

Some of the key conclusions that can be drawn from this research are as follows:

- French, German and British consumers are receptive to the concept of foods with benefits over and above that of pure nutrition;
- Not understanding the expression functional foods, most consumers perceive such products as just another 'healthier' alternative to standard foodstuffs;
- For manufacturers to compete successfully on the European stage, they will need to be aware of the different national attitudes and perceptions to diet and well-being;
- Health concerns for self are different to those for a partner and for children;
- Familiarity of well-known brands/companies does seem to add some reassurance to consumers when faced with new products and concepts;

Finally, I think it is important to stress that while traditionally enriched foods and drinks, functional foods and supplements are competing for the same share of the consumer purse, in reality it is not that simple. Clearly, these products are not mutually exchangeable in the consumer's eyes and, as a consequence, the winners of tomorrow will be those companies that best understand how consumers mix and match these product categories to achieve perceived levels of health control.

Bovine milk: A Unique Source of Immunomodulatory Ingredients for Functional Foods

H.S. Gill, K.J. Rutherfurd and M.L. Cross

MILK AND HEALTH RESEARCH CENTRE, INSTITUTE OF FOOD, NUTRITION AND HUMAN HEALTH, MASSEY UNIVERSITY, PRIVATE BAG 11222, PALMERSTON NORTH, NEW ZEALAND

1 Introduction

Consumer demands for functional foods have increased dramatically over the last decade. In particular, foods that offer enhanced health benefits against disease are becoming increasingly important, especially among consumer groups who may be at greater risk of developing disease, such as the young, the elderly and immunocompromised individuals.

Bovine milk has long been recognized as an important source of protein and fats for human consumption. However, it is only recently that the potential for milk to serve as a functional food, *i.e.* one which imparts a measurable health benefit to the recipient, has been acknowledged.[1] Accordingly, the last decade has seen an increasing amount of research into the testing of milk components and fractions for their potential to influence human health in relation to the control of disease.

It is now evident that bovine milk contains a number of biologically active components that may potentially be incorporated into health-promoting functional foods for human consumption. The majority of the research in this field has sought to identify and characterize components that influence the immune system, and how this may affect an individual's resistance to disease. The term *immunomodulation* has been applied to milk components that have been shown to influence immune function, such that this influence may promote enhanced disease resistance.

This review outlines the most recent scientific evidence for the presence of immunomodulatory components in bovine milk, and discusses the potential of these components to be incorporated as immune-enhancing ingredients into functional foods with defined health-promoting qualities.

2 Milk as a Biologically Active Fluid

The idea that milk is capable of modulating physiological function, in addition to providing nutrition, is not a recent concept.[2] However, significant progress in this line of research has been made over the past decade by an improved understanding of the immunological role of milk *in vivo* in its natural producer, *i.e.* the dairy cow. The neonatal bovine enters the world with poorly developed immunity and maternal milk provides a rich source of molecules which are able to modulate the developing immune system, so that it becomes fully functional.[3] This is particularly important with respect to the establishment of protective immunity against infectious diseases as the newborn's environmental exposure to potentially dangerous pathogens increases, but equally important in ensuring that the developing immune system does not become overactive, which would lead to aggressive and unwanted immunological responses such as atopic reactions.

Furthermore, because there is very little transfer of maternal immunoglobulin across the placenta, the neonatal bovine has a circulatory system deficient of protective antibodies. This is compensated by an extremely high concentration of antibody molecules (predominantly IgG1 class) in the colostrum and post-partum milk. These antibodies provide ready-made passive protection to the newborn against infectious agents.

Taken together, this presents a picture of bovine milk as a biologically active fluid, capable of modulating the developing immune system of the newborn and simultaneously providing it with passive protection against pathogens. Not surprisingly, this recent understanding has led nutritionists and immunologists to research the precise biological role of milk-derived sources of immunomodulatory molecules, in order that they might be exploited into functional food products with proven health-promoting qualities.

3 Milk-derived Sources of Immunomodulatory Molecules

During the commercial manufacturing of milk products, milk is first skimmed to remove fats and the residual protein is fractionated into caseins and whey. The majority of immunomodulatory activity of bovine milk has been detected in sub-fractions and enzymatic digests of the major milk proteins (caseins and whey), although recent evidence has also suggested that components of milk fat can modulate immune function (Table 1). The vast majority of research that has sought to characterize the immunomodulatory effects of milk components has been conducted on *ex vivo* aspects of the immune system. In the majority of cases, this has involved the isolation of immune effectors (*e.g.* leucocytes) from laboratory animals (usually rodents), and the investigation of how immune function may be affected *in vitro* by exposure to milk components. Rather less attention has been paid to how dietary milk components might affect immune function in laboratory animals *in vivo* and, as will be discussed, even fewer investigations have expanded this point to include clinical studies on the influence of dietary milk components on human immune function.

4 The Ability of Milk to Modulate Immune Function

4.1 Animal Models

Several *in vitro* studies have indicated that milk components have the potential to modulate immune function. In most cases, modulation has been recorded following culture of leucocytes (obtained *ex vivo* from rodents) with a defined milk component, and then subsequently measuring how the immune function has been affected. The majority of this line of research has focussed on the *in vitro* function of lymphocytes (*e.g.* cellular proliferation and cytokine production) (Table 1).

This latter point has highlighted one of the major limitations to using an *in vitro* approach to identify immunomodulatory activity in milk, *i.e.* that exposure to gastrointestinal enzymes can radically alter the observed activity. This raises the question as to whether the immunomodulatory activity described by the native molecule takes place *in vivo* or not.[12] Consequently, research over the past few years has focussed increasingly on *in vivo* dietary trials of milk components in laboratory animal models, and the measurement of their effects on immune function. Research by Bounous and colleagues[13,14] utilized this approach in the early 1980s, and demonstrated that the whey protein α-lactalbumin could dramatically enhance lymphocyte function when included in a dietary formulation fed to mice (even though no previous or subsequent studies have conclusively demonstrated an immunomodulatory effect of α-lactalbumin on murine lymphocytes *in vitro*).

Further studies in mice have confirmed that both dietary α-lactalbumin and whole whey protein concentrate (WPC) can enhance the responsiveness of spleen-derived lymphocytes to T and B cell mitogens.[15,16] The immune enhancing effect of α-lactalbumin was also demonstrated when mice were fed enzyme-derived hydrolysates of the protein, and the immunomodulatory effect increased with increasing dietary levels of the hydrolysates.[17] Subsequently, both lactoferrin and κ-casein derived caseinoglycopeptide (CGP) have been shown to enhance lymphocyte function in mice following dietary inclusion;[18,19] it is noteworthy that both of these components had been demonstrated previously in *in vitro* studies to suppress murine lymphocyte function.

While milk-derived proteins and peptides have received the majority of research attention, there is evidence that milk fat-derived components can also modulate immune function. In particular, dietary inclusion of the fatty acid conjugated linoleic acid (CLA) has been shown to enhance *in vitro* lymphocyte proliferation in mice (Gill *et al.*, unpublished), as well as enhance *in vivo* lymphocyte infiltration responses (delayed type hypersensitivity test) in rats.[20]

Table 1 *Summary of modulatory effects on lymphocytes by individual milk components and their enzyme-generated peptide fractions*

	Sub-fraction	Effect on immune function	
		Native state	Enzyme digests
Milk proteins			
Casein	α-Casein	Enhances proliferation of lymphocytes *in vitro*	Enhances lymphocyte proliferation
		Suppresses proliferation of lymphocytes (pancreatin, trypsin, and pepsin/trypsin)	
	β-Casein	Enhances proliferation of lymphocytes *in vitro*	Suppresses proliferation of lymphocytes (pancreatin, trypsin, and pepsin/trypsin)
			Enhances lymphocyte proliferation (pepsin/chymosin)
	κ-Casein	Suppresses lymphocyte proliferation *in vitro*	Suppresses proliferation of lymphocytes (chymosin)
			Enhances proliferation of lymphocytes (trypsin and pepsin/trypsin)
Whey	Whey Protein Concentrate	Enhances lymphocyte proliferation *in vivo*; suppresses lymphocyte proliferation *in vitro*	No data
	α-Lactalbumin	Enhances proliferation of lymphocytes *in vivo*	Hydrolysed lactalbumin stimulates B lymphocytes
	Lactoferrin	Suppresses lymphocyte proliferation *in vitro*	Modulates lymphocyte proliferation *in vitro*
	Lactoperoxidase	Suppresses lymphocyte proliferation *in vitro*	No data
	Milk Growth Factor	Suppresses lymphocyte proliferation *in vitro*	N/A
Milk fats			
Fatty Acids	Conjugated Linoleic Acid (CLA)	Enhances lymphocyte proliferation *in vivo*	N/A

4.1.1 Effects of Bovine Milk Components on Antibody Responses. In addition to affecting lymphocyte function, dietary milk proteins have been shown to modulate antibody responses in animal models. α-Lactalbumin, α-lactalbumin hydrolysates and whole whey protein concentrate have each been shown to enhance antibody production to systemically-delivered foreign antigens in mice.[13–17] In contrast, dietary caseinoglycopeptide has been shown to suppress serum antibody responses against both orally- and systemically-administered foreign antigens in mice.[19]

Although antibody responses are generally considered to be beneficial to host immunity, the development of IgE antibody responses can lead to allergic hypersensitivity reactions. Watson *et al.*[21] showed that mice fed a colostral whey-derived extract produced lower IgE antibody responses to foreign antigen. Further research has demonstrated that both κ-casein and lactoferrin can suppress IgE-mediated hypersensitivity responses in guinea pigs *in vivo*, possibly by inhibiting histamine release.[22,23]

4.1.2 Effects of Bovine Milk Components on Non-Lymphoid Immune Functions. In common with the research on lymphoid cell function, the majority of research in this field has focused on the ability of milk components to modulate immune function *in vitro*. α- and κ-casein, as well as whey-derived lactoferrin, have been shown to suppress *in vitro* phagocytic responses by murine macrophages.[24,25] In contrast, β-casein and α-lactalbumin were shown to enhance macrophage function.[5,26] Furthermore, β-casein and whole whey protein concentrate were able to enhance the function of ovine neutrophils when these proteins were included in *in vitro* cell culture.[5,26]

4.2 Human Studies

Several *in vitro* studies have indicated that the human immune system can be affected by bovine milk-derived components. Peptides derived from the enzymatic cleavage of α- and β-casein have been shown to suppress mitogen-stimulated proliferation of human blood- and lamina propria-derived lymphocytes *in vitro*,[27–30] and the whey derived fraction milk growth factor (MGF) has been shown to be a potent suppressive factor for human lymphocyte function.[31] In contrast, Sutas *et al.*[32] have shown that peptides derived from the enzymatic cleavage of κ-casein can enhance human lymphocyte function *in vitro*. Intact bovine lactoferrin (and its pepsin-cleavage product lactoferricin) have both been shown to enhance human neutrophil function in *in vitro* cell culture.

In general, there is a lack of clinical studies that have sought to measure the immunomodulatory effects of dietary bovine milk-derived components on human immune function. However, Brosche and Platt[33] have shown that elderly subjects had enhanced blood cell phagocytic function following consumption of a milk-based diet. Recent research by our group has also confirmed that dietary milk can enhance human phagocytic cell function.[34]

5 Enhanced Resistance to Disease

5.1 Enhanced Resistance *via* Immune Modulation

Bounous and colleagues have demonstrated enhanced lymphocyte function and antibody responses in mice fed increasing concentrations of α-lactalbumin hydrolysates, and this heightened immunity corresponded with enhanced resistance to challenge with *Salmonella typhimurium*[17] (although whether this was a causal relationship was not investigated). Protective effects and immune enhancement have been shown for other milk-derived peptides against pathogen challenge in mice;[35,36] however, in these cases the bioactive peptides were administered by systemic injection prior to pathogen challenge, and so their potential role in dietary inclusion remains unconfirmed.

In addition to providing enhanced resistance against pathogens, milk-derived components have also been shown to promote immune responses that are directly relevant to the immune-mediated control of pathogenic effects. Recent research has shown that dietary whey protein concentrate can enhance intestinal mucosal antibody responses against orally administered cholera toxin and tetanus toxoid vaccines in mice.[37]

Many other health-promoting effects have been demonstrated for milk products. However, in the majority of cases there has been no conclusive evidence that the proven health benefit was derived from an enhancement of immune function. Bounous and colleagues have demonstrated that dietary whey proteins can enhance resistance to tumours in mice,[38] and in addition can promote overall weight gain in HIV-positive men.[39] Dietary whey proteins have been shown by the same authors to increase tissue concentrations of glutathione,[40] and it may be that this promotes enhanced cellular immune function. Neurath *et al.*[41] have demonstrated a more direct beneficial effect of whey proteins on HIV-related health, in that a modified α-lactoglobin preparation was shown to inhibit viral binding to cell surface CD4 receptors on T lymphocytes *in vitro*.

5.2 Enhanced Resistance *via* Passive Transfer of Immunoglobulins

Since bovine milk (particularly post-partum milk and colostrum) contains a high concentration of immunoglobulins,[3] there has been extensive research into the potential benefits of immunoglobulin-rich products in human health. The protective effect is mediated by specific binding of immunoglobulins to pathogens in the gastrointestinal tract. Since lactating cows are likely to have been exposed to several intestinal pathogens during their lifetime, pathogen-specific antibodies may be present in their mammary secretions which could offer passive protection to humans. Indeed, Rump *et al.*[42] have demonstrated that the incidence of chronic diarrhoea among HIV-infected individuals can be significantly reduced, following consumption of a diet containing bovine colostral immunoglobulin concentrate.

While the use of high immunoglobulin content colostrum from normal cows offers potential for the development of pathogen-limiting therapeutics, most development in this area has focussed on the use of immunoglobulin-containing colostral products from cows hyperimmunised against specific pathogens. Such hyperimmune colostrum has a far higher concentration (titre) of pathogen-specific immunoglobulins. Controlled clinical trials have also been carried out using hyperimmune bovine colostrum, containing specific antibodies against *Shigella flexneri, Helicobacter pylori, Vibrio cholerae,* caries streptococci or rotavirus.[43–47] While results have been variable, the evidence has generally indicated that colostrum preparations, with high antibody titres against defined pathogens, can provide significant protection against gastrointestinal disease in humans. In particular, the use of hyper-immune colostrum, containing high titres of antibody specific to rotavirus, has been shown to be effective in the treatment of infant diarrhoea.[48] These studies open up the way for the development of immunoglobulin-containing milk or colostrum preparations into marketable dairy foods with defined health benefits. However, the industry must be aware of the continued need for safety monitoring of bovine products, and must also be sensitive to the ethical considerations of consumers when promoting products derived from hyper-immunized cows.

6 Conclusions

Bovine milk contains a host of immunomodulatory molecules that could be incorporated into health-promoting functional foods. Such foods are likely to be of particular benefit to subjects with compromised or poorly functioning immune systems, such as the young, elderly people or immunodeficient individuals. Immunomodulatory molecules from milk could also be incorporated into functional foods that aim to control undesirable consequences of an over-active immune system, such as chronic intestinal inflammation or atopic hypersensitivity reactions. In addition to promoting improvements in active immune function, bovine milk also contains pathogen-specific immunoglobulins that can be used to provide passive immune protection to at-risk individuals. In particular, the use of hyperimmune colostrum-based products offers great promise to combat gastrointestinal pathogens and alleviate diarrhoea.

There is plenty of evidence to indicate that dietary milk components can modulate immune function *in vivo*; however, the possibility still exists that subtle influences on the immune system may remain undetected, highlighting the need for further research in this area. The most recent progress toward the development of functional foods has come from animal studies which have shown that health-related immune functions (*e.g.* responses to vaccination, and protective responses against pathogens and tumours) can be directly enhanced by dietary milk products. Further studies of this kind are necessary to confirm that the ability to modulate immune function corresponds with a measurable improvement in health. Finally, there is an on-going need to verify the efficacy and safety of milk-based functional foods in controlled clinical trials.

References

1   H. Korhonen, A. Pihlanto-Leppala, P. Rantamaki and T. Tupasela, *Agric. Food Sci. Finland*, 1998, **7**, 283.
2   T.J. Newby, C.R. Stokes and F.J. Bourne, *Vet. Immunol. Immunopathol.*, 1982, **3**, 67.
3   B.M.L. Goddeeris and W.I. Morrison, 'Cell mediated immunity in ruminants', CRC Press, London, 1994.
4   R.I. Carr, *Ann. NY Acad. Sci.*, 1990, **574**, 374.
5   C.W. Wong, H.F. Seow, A.J. Husband, G.O. Regester and D.L. Watson, *Vet. Immunol. Immunopathol.*, 1997, **56**, 85.
6   H. Otani and I. Hata, *J. Dairy Res.*, 1995, **62**, 339.
7   P.M. Torre and S.P. Oliver, *J. Dairy Sci.*, 1989; **72**: 219–27.
8   O. Barta, V.D. Barta, M.V. Crisman, R.M. Akers, *Am. J. Vet. Res.*, 1991, **52**, 247.
9   J.J. Rejman, K.D. Payne, M.J. Lewis, P.M. Torre, R.A. Muenchen and S.P. Oliver, *Food Agric. Immunol.*, 1992, **4**, 253.
10 H. Otani and M. Odashima, *Food Agric. Immunol.*, 1997, **9**, 193.
11 H. Otani and M. Monnai, *Food Agric. Immunol.*, 1993. **5**, 219.
12 H. Meisel, *Biopol.*, 1997, **43**, 119.
13 G. Bounous and P.A.L. Kongshavn, *J. Nutr.*, 1982, **112**, 1747.
14 G. Bounous, L. Letourneau and P.A.L. Kongshavn,. *J. Nutr.*, 1983, **113**, 1415.
15 G. Bounous, P.A.L. Kongshavn and P. Gold, *Clin. Invest. Med.*, 1988, **11**, 271.
16 C.W. Wong and D.L. Watson, *J. Dairy Res.*, 1995, **62**, 359.
17 G. Bounous, M.M. Stevenson and P.A.L. Kongshavn, *J. Infect. Dis.*, 1981, **144**, 281.
18 H. Debabbi, M. Dubarry, M. Ranteau and D. Tome, *J. Dairy Res.*, 1998, **65**, 283.
19 M. Monnai, Y. Horimoto and H. Otani, *Milchwissenshaft*, 1998, **53**, 129.
20 B.P Chew, T.S. Wong, T.D. Shultz and N.S. Magnuson, *Anticancer Res.*, 1997, **17**, 1099.
21 D.L. Watson, G.L. Francis and F.J. Ballard. *J. Dairy Res.*, 1992, **59**, 369.
22 H. Otani and Y. Yamada, *Milchwissenschaft*, 1994, **49**, 20.
23 H. Otani and Y. Yamada, *Milchwissenschaft*, 1995, **50**, 549.
24 H. Otani and M. Futakami, *Anim. Sci. Tech.*, 1994, **65**, 423.
25 H. Otani and M. Futakami, *Food Agric. Immunol.*, 1996, **8**, 59.
26 C.W. Wong, H.F. Seow, A.H. Liu, A. Husband, G.W. Smithers and D.L. Watson, *Immunol. Cell Biol.*, 1996, **74**, 323.
27 H. Kayser and H. Meisel, *FEBS Lett.*, 1996, **383**, 18.
28 E. Schlimme and H. Meisel, *Die Nahrung*, 1995, **39**, 1.
29 E. Lahov and W. Regelson, *Food Chem. Tox.*, 1996, **34**, 131.
30 Y. Elitsur and G. D. Luk, *Clin. Exp. Immunol.*, 1991, **85**, 493.
31 M. Stoeck, C. Ruegg, S. Miescher, S. Carrel, D. Cox, V. Fliedner, S. Alkan and V. VonFliedner, *J. Immunol*,. 1989, **143**, 3258.
32 Y. Sutas, E. Soppi, H. Korhonen, E.L. Syvaoja, M. Saxelin, T. Rokka and E. Isolauri, *J. Allerg. Clin. Immunol.*, 1996, **98**, 216.
33 T. Brosche and D. Platt, *Ageing: Immunol. Infect. Dis.*, 1995, **6**, 29.
34 R.K. Chandra and H.S. Gill, *Eur. J. Clin. Nutr.* (manuscript in preparation).
35 P. Jolles, A.M. Fiat, D. Migliore-Samour, L. Drouet and J.P. Caen, 'New Perspectives in Infant Nutrition', B. Renner and G. Sawatzki (eds.), Thieme Med. Publ., New York, 1992, p. 160.

36   K. Shimizu, H. Matsuzawa, K.Okada, S. Tazume, S. Dosako, Y. Kawasaki, K. Hashimoto and Y. Koga, *Arch. Virol.*, 1996, **141**, 1875.

37   H.S. Gill and K.J. Rutherfurd, *Bull. Int. Dairy Fed.*, 1998, **336**, 31.

38   G. Bounous, R. Papenburg, P.A.L. Kongshavn, P. Gold and D. Fleiszer, *Clin. Invest. Med.*, 1988, **11**, 213.

39   G. Bounous, S. Baruchel, J. Falutz and P. Gold, *Clin. Invest. Med.*, 1993, **16**, 204.

40   G. Bounous and P. Gold, *Clin. Invest. Med.*, 1991, **14**, 296.

41   A.R. Neurath, Y.Y. Li, N. Strick and S. Jiang, *Lancet*, 1996, **347**, 1703.

42   J.A. Rump, R. Arndt, A. Arnold, C. Bendick, H. Dichtelmueller, M. Franke, E.B. Helm, H. Jaeger, B. Kampmann *et al.*, *Clin. Invest.*, 1992, **70**, 588.

43   C.O. Tacket, S.B. Binion, E. Bostwick, G. Losonsky, M.J. Roy and R. Edelman, *Am. J. Tropical Med. Hygien.*, 1992, **47**, 276.

44   T.H. Casswall, S.A. Sarker, M.J. Albert, G.J. Fuchs, M. Bergstrom, L. Bjorck and L. Hammarstrom, *Aliment. Pharmacol. Theraput.*, 1998, **12**, 563.

45   M. Boesman-Finkelstein, N.E. Walton and R.A. Finkelstein, *Infect. Immun.*, 1989, **57**, 1227.

46   S.M. Michalek, J.R. McGhee, R.R. Arnold and J. Mestecky, *Adv. Exp. Med. Bio.*, 1978, **107**, 261.

47   A.K. Mitra, D. Mahalanabis, H. Ashraf, I. Unicomb, R. Eeckels and S. Tzipori, *Acta Pediatr.*, 1995, **84**, 996.

48   T. Ebina, A. Sato, K. Umezu, N. Ishida, S. Ohyama, A. Oizumi, K. Aikawa, S. Katagiri, N. Katsushima, A. Imai *et al.*, *Med. Microbiol. Immunol.*, 1985, **174**, 177.

Catabolite Regulation: An Intrinsic Role for Fructo-Oligosaccharides?

B.A. Rabiu, A.L. McCartney, G.T. Macfarlane* and
G.R. Gibson

DEPARTMENT OF FOOD SCIENCE AND TECHNOLOGY,
UNIVERSITY OF READING, READING RG6 6AP, UK
* MEDICAL SCHOOL, NINE WELLS HOSPITAL, UNIVERSITY OF
DUNDEE, SCOTLAND, UK

1 Introduction

The human colon is a multisubstrate environment that supports a highly diverse community of microorganisms.[1] Some groups, such as the bifidobacteria and lactobacilli, contain species which are loosely termed as beneficial.[2] As a result, there is much interest in modulation of such bacterial populations that are resident to the colon. One approach uses prebiotics and is founded on the concept that certain non-digestible short-chain carbohydrates (SCCs), for example fructooligosaccharides (FOS), are selectively fermented by certain bacterial populations in the colon.[3,4] Studies with volunteers, monitoring the prebiotic potentials of FOS,[3,5] have demonstrated that health consequences are dependent on the type of prebiotic used, dose ingested and the indigenous bacteria of the host.

In the current study, we have investigated the functional attributes of FOS by examining its metabolism in the presence of starch, another polymerised carbon source (PCS) present in the diet, using an intestinal bifidobacterial isolate.

2 Materials and Methods

2.1 Culture Conditions

Pure culture investigations were carried out in single stage continuous culture chemostats with *Bifidobacterium bifidum* (strain BS4, faecal isolate), at dilution rates of 0.03 h^{-1} and 0.15 h^{-1} (pH 6.2, 37 °C).[6] Starch (10 g l^{-1}) was used as the sole carbohydrate source, in a nutrient medium reservoir, serving as a 'control' to an identical 'test' system containing starch and FOS (10 g l^{-1} each). Once steady state conditions had been reached (at least seven turnovers in the system) 20 ml aliquots were removed for analysis of enzymes involved in hydrolysis of the carbohydrates.[7] One millilitre was used for the estimation of residual carbohydrates.[8] Replicate assays were carried out for each determination.

Batch culture experiments were also performed, in which *B. bifidum* enriched on starch (as the sole carbohydrate source) was spiked with FOS (1% w/v). Samples were removed at periodic intervals and treated as above. Bacteriological enumeration was also carried out at each time point using Wilkins Chalgren agar.

2.2 Carbohydrate Analysis

The residual carbohydrate in the spent medium from the fermentation vessels was determined as alditol acetates using gas chromatography.[8]

2.3 Enzyme Assays

Supernatant fractions of chemostat effluent (15 ml) were prepared by centrifugation (20000 *g*, 30 min, 4 °C). Crude cell extracts were obtained using a French pressure cell (1.1 × 10^5 KPa). α-Amylase and β-2,1-fructofuranosidase (Ffase) activities were determined by following the release of reducing sugars from starch and FOS respectively, employing a dinitrosalicylic acid reagent.[7] The starch hydrolysing enzyme, α-glucosidase, was assayed by using its *p*-nitrophenyl substrate.[7]

3 Results and Discussion

3.1 Characteristics of the Continuous Culture Systems

Table 1 shows how steady state conditions were reached in the chemostats.[9] The presence of FOS in the 'test' vessels increased the fermentation rate by 31% and 6% at low and high dilution rates, respectively. Increased nutrient availability stimulated fermentation by at least one order of magnitude, as compared to the carbon-limited chemostats.

3.2 Effect of FOS on Hydrolytic Enzyme Activities

Results in Figure 1 show polysaccharidase activities in the fermentation systems. These data indicate that Ffase activity was induced by the presence of FOS (19.4%), and this was associated with a reduction in the activity of the starch-degrading enzymes, amylase and α-glucosidase (2.9% and 15.8% respectively, at a dilution rate of 0.03 h^{-1}; and 45.9% and 15.3% respectively, at the

Table 1   *Characteristics of the continuous culture systems*

	Dilution rate (h^{-1})	0.03	0.15
Control	D_{alk} (h^{-1})	0.004	0.02
	W $(mg\ ml^{-1})$	1.45	3.00
	$^*Q_{alk}$	45.98 ± 0.01	111.1 ± 0.07
Test	D_{alk} (h^{-1})	0.008	0.026
	W $(mg\ ml^{-1})$	2.22	3.69
	$^*Q_{alk}$	60.06 ± 0.07	117.4 ± 0.05

Results are means $\pm$SEM $(n=3)$

*nmol OH⁻ utilised min^{-1} (mg dry wt culture)$^{-1}$ = $(D_{alk}$ x A)$/ (W$ x 60), where D_{alk} = dilution rate for alkali [vol. of alkali added h^{-1}(vol. of culture vessel)$^{-1}$], A = concentration of alkali [nmol ml^{-1}], W = community dry weight [mg ml^{-1}]

higher dilution rate). Increasing dilution rate from $0.03\ h^{-1}$ to $0.15\ h^{-1}$ was shown to reduce Ffase activity in both test and control systems. Additionally the reduction of amylase activity was markedly greater at the higher dilution rate. Moreover, analyses of residual starch concentrations showed that FOS reduced the amount of starch metabolised by 20% at the higher dilution rate and by 2% at $0.03\ h^{-1}$.

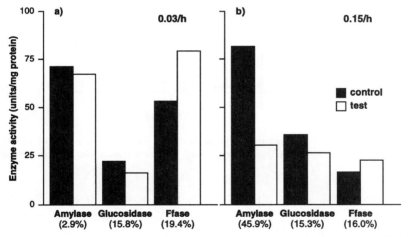

Figure 1   *Effect of FOS on starch depolymerising enzymes in pure cultures of B. bifidum at dilution rates of (a) 0.03 h^{-1} and (b) 0.15 h^{-1}. Results are means of two separate enzyme assays ($\pm$SD < 0.04). Results in parentheses are % differences between control and test*

3.3 Effect of FOS on Starch Utilisation in Batch Cultures

Enzyme measurements for the fermentation of FOS by starch-adapted pure cultures of *B. bifidum*, showed that starch degradation ceased when FOS was introduced. There was a concurrent reduction in α-amylase and α-glucosidase activities by 41% and 50%, respectively (Figure 2), suggesting catabolite regulation of these enzymes. This was accompanied by an increase in Ffase activity (Figure 2), and moderate increases in viable bacterial counts (Figure 3).

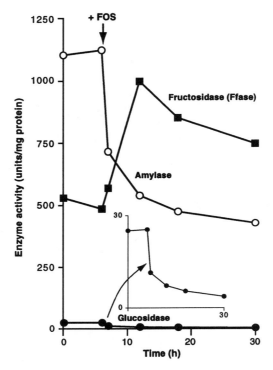

Figure 2 *Effect of FOS dose on enzyme activities in pure cultures of starch adapted B. bifidum cells. Data points are means of two separate enzyme determinations (±SD < 0.21)*

The mechanism assumed here is catabolite inhibition of starch depolymerising enzymes by FOS. Catabolite repression was ruled out as enzymes synthesised before introduction of a preferred substrate would still be active, thus masking any immediate responses. Additionally, the fermentation pattern here was not consistent with simple competition for transport, when a fixed ratio of FOS *versus* starch utilisation would be expected. [10]

Substrate utilisation was determined by analysis of residual carbohydrate monomers. Results in Figure 4 show that an increase in FOS in the culture media was associated with a near zero rate of starch depolymerisation, consistent with observed reductions in polysaccharidase activities. However, this repression appeared to be reversible, as during a reduced rate of FOS hydrolysis, starch depolymerisation was active (Figure 4).

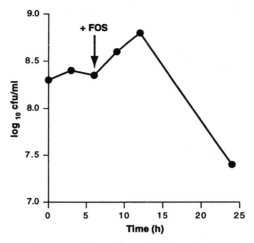

Figure 3 *Effect of FOS dose on numbers of starch adapted B.bifidum cells. Data points are means of two separate plate counts (±SD < 0.21)*

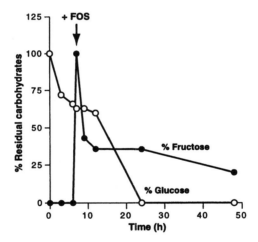

Figure 4 *Effect of FOS on the utilisation of starch by starch adapted B. bifidum cells. Residual glucose and fructose monomers determined after acid hydrolysis of starch and FOS respectively.[8] Data points are means of two determinations (±SD < 0.11)*

4 Conclusion

Results here have indicated that selective degradation of FOS by bifidobacteria may have an important role in regulating levels/rates of enzymic utilisation of other polymerised carbon sources. As such, catabolite regulation may allow for increased carbohydrate fermentation in more substrate limited areas of the colon, *e.g.* distal regions. However, further confirmatory work would need to be carried out using more test species and substrates in a competitive microbial ecosystem which more resembles the complexity of the colon.

References

1   G.T. Macfarlane and J.H. Cummings, The colonic flora, fermentation and large bowel digestive function. In *The Large Intestine: Physiology, Pathophysiology and Disease*, pp. 51–92, eds. Phillips S.F. *et. al.* New York: Raven Press, 1991.

2   R. Fuller and G.R. Gibson, *Clin. Microbiol. Infect.*, 1998, **4**, 477.

3   G.R. Gibson, E.B. Beatty, X. Wang and J.H. Cummings, *Gastroenterology*, 1995, **108**, 975.

4   G.R. Gibson, *Brit. J. Nutr.*, 1998, **80**, S209.

5   M.B. Roberfroid, J.A.E. Van Loo and G.R. Gibson, *J. Nutr.*, 1998, **128**, 11.

6   G.T. Macfarlane and H.N. Englyst, *J. Appl. Bacteriol.*, 1986, **60**, 195.

7   B.A. Degnan and G.T. Macfarlane, *Current Microbiol.*, 1994, **29**, 43.

8   H.N. Englyst, M.E. Quigley and G.J. Hudson, *Analyst*, 1994, **119**, 1497.

9   A.S. Mckee, A.S. McDermid, D.C. Ellwood and P.D. Marsh, *J. Appl. Bacteriol.*, 1985, **59**, 263.

10   J.B. Russell and R.L. Baldwin, *Appl. Environ. Microbiol.*, 1978, **36**, 319.

Probiotic Bifidobacteria and Their Identification Using Molecular Genetic Techniques

S. McBrearty[1], P.J. Simpson[1], G. Fitzgerald[2], J.K. Collins[2], R. P. Ross[1] and C. Stanton[1]

[1] TEAGASC, DAIRY PRODUCTS RESEARCH CENTRE, MOOREPARK, FERMOY, CO. CORK, IRELAND
[2] DEPARTMENT OF MICROBIOLOGY, UNIVERSITY COLLEGE CORK, IRELAND

1 Introduction

The ingestion of high levels of probiotic bacteria has been associated with a variety of health benefits. These include nutritional benefits such as enhanced bioavailability of minerals and synthesis of vitamins, and therapeutic benefits such as alleviation of lactose intolerance and suppression of cancer.[1]

Most probiotic strains studied or utilised to date belong either to the genera of *Lactobacillus* or *Bifidobacterium*[2] and in recent years there has been a rapid development of molecular genetic methods to identify and characterise the bacteria of both genera.[3] Studies relating solely to bifidobacteria are considered in this article. The majority of research has used the discriminatory powers of molecular techniques to differentiate specific *Bifidobacterium* strains from the large number of closely related strains present in the gastrointestinal tract (GIT).[4]

The enumeration of probiotics from food products is traditionally achieved through dilution plating on either selective or non-selective media.[5-9] However, several factors affect the ability to recover and enumerate bifidobacteria using this approach. These include the particular *Bifidobacterium* strain, the nature of the food product and the type of selective media used.[10-13] Molecular genetic techniques have the capacity to screen large numbers of colonies to determine the taxonomic origin of the isolates and, therefore, reduce the need to selectively inhibit non-bifidobacteria.[14-16] However, to date it appears that only one study has used molecular probes to screen colonies and enumerate bifidobacteria in food products.[15] Moreover, techniques developed to detect specific bifidobacteria directly from faecal samples without cell culturing[17-19] have yet to be applied to probiotic food products.

Numerous commercial food products containing bifidobacteria are currently available and the list continues to expand[20] (Table 1). However, given the fastidious and anaerobic nature of bifidobacteria,[21] the development of food products containing these strains at the recommended level of 10^7 cfu/ml or g of product[22] can be particularly challenging.[6,11,20] For example, four retail probiotic yoghurts tested in our laboratory contained between 10^5 and 10^7 cfu/ml of bifidobacteria.

Table 1   *Examples of probiotic food products containing bifidobacteria*

Product	Bifidobacterium strains	Viable cells in product (cfu/g or ml)	Refs
Acidophilus bifidus yoghurt	*B. bifidum* and *B. longum*	$1-3 \times 10^7$	49
Greek-style yoghurt	*B. bifidum*	$4.6 \times 10^5 - 4.1 \times 10^7$	50
Fermented milk (Olifus®)	*B. bifidum*	10^8	49
Cheddar cheese	*B. bifidum*	$1.4 \times 10^7 - 2.6 \times 10^7$	51
Gouda cheese	*Bifidobacterium* sp. strain Bo	$6 \times 10^8 - 1.8 \times 10^9$	52
White-brined cheese	*B. adolescentis* and *B. bifidum*	$3.4 \times 10^3 - 1.96 \times 10^4$ and $2.60 \times 10^4 - 5.50 \times 10^5$	53
Crescenza cheese	*B. bifidum, B. longum* and *B. infantis*	$1.12 \times 10^8, 1.32 \times 10^7$ and 1.69×10^5	54
Cultured cottage cheese	*B. infantis*	$1.0 \times 10^8 - 6.3 \times 10^8$	55
Fresh cheese	*B. breve* and *B. longum*	$> 10^6$ (after 15 days)	56
Goats cheese	*B. lactis*	7.5×10^6	57
Mayonaisse	*B. bifidum* and *B. infantis*	1×10^4 and 1×10^5	58
Ice cream	*B. bifidum*	1×10^7	59

This communication will review the current molecular genetic techniques used to detect and enumerate bifidobacteria in faecal samples and probiotic food products. In addition, we present data on the use of random amplified polymorphic DNA (RAPD) PCR to characterise bifidobacteria obtained from culture collections, commercial sources and the human GIT.

2 Molecular Genetic Methods to Identify Bifidobacteria

2.1 Pulse-Field-Gel-Electrophoresis (PFGE)

If restriction endonucleases cut at specific DNA sequences that occur infrequently within the genome, a discrete number of DNA fragments can be generated. In bacteria with a G + C content above 45% (bifidobacteria contain between 57 and 64%), the rarity of the tetranucleotide CTAG means that restriction endonucleases such as *Xba*I and *Spe*I will cut the genome infrequently and generate between 10 and 30 DNA fragments ranging from 20 to 400 kb.[23] The relatively large DNA fragments can be resolved by continually switching the direction of DNA migration within an agarose gel. PFGE achieves this by changing the direction of the electric field. The resulting gels generate a restriction fragment length polymorphism (RFLP) patterns, which reflect differences in DNA sequence at the restriction enzyme recognition site. Using such RFLP patterns, a range of bifidobacteria from commercial probiotic preparations, culture collections and human isolates have been distinguished.[23-26] The RFLP fingerprints can not be used to speciate strains because intraspecies and interspecies differences are comparable. In addition, the time and labour demands associated with PFGE make routine analysis of large numbers of isolates impractical.

2.2 DNA-DNA and DNA-RNA Hybridisations

Bifidobacteria at the genus, species and strain levels can be identified on the basis of their genetic variation using nucleic acid hybridisation. Probes can be applied either as labelled probes, which hybridise to target DNA or ribosomal RNA, or as primers to DNA sequences to initiate a polymerase chain reaction (PCR).

2.2.1 rRNA Probes and Ribotyping. Ribotyping involves restricting the total DNA of a bacterium with a restriction endonuclease followed by electrophoresis, transfer of DNA onto filters and hybridisation with labelled probes that target the ribosomal genes of any bacteria. As there are generally multiple copies of the ribosomal RNA genes present in a bacterial cell, a number of different sized fragments will hybridise to the probe, giving a characteristic pattern.[27]

Using part of the 23S rRNA gene as a probe and several restriction endonucleases, ribotyping was able to differentiate a wide range of bifidobacteria strains from culture collections, industrial sources and human faecal isolates.[28,29] Common bands evident in the culture collection strains allowed for a tentative speciation of several of the human isolates.[28] In the same study, human subjects receiving antibiotics were also fed an antibiotic resistant *Bifidobacterium* strain within a fermented dairy product and ribotyping was used to identify changes in their microflora. Antibiotic treatment led to a change in the dominant bifidobacteria, which persisted for the two month period of analysis. The ingested strain was present at $6-7 \times 10^8$ cfu/g of faeces during the period of consumption, which represented 30% of the oral load, but rapidly disappeared following cessation of the ingestion period.

In another study, ribotyping differentiated bifidobacteria isolated from human faeces.[25] Subribotypes were noted where the degree of difference was not considered to be sufficient to represent a new strain. PFGE analysis was also performed and, overall, both techniques concurred well. However, some discrepancies were noted, for example some strains differentiated by ribotyping had identical PFGE patterns and in other cases strains belonging to the same subribotype had different PFGE patterns. In a separate study, two closely related *B. breve* strains that were distinguishable by ribotyping had identical PFGE profiles.[24]

In general, ribotyping involves a considerable amount of sample manipulation, which may include the handling of radioactive material, and this limits its application to the routine screening of large numbers of colonies.

2.2.2 Specific rRNA Probes. In recent years, 16S rRNA probe hybridisation has become widely adopted for the detection of specific bacterial groups in mixed populations.[30,31] This methodology uses synthetic oligodeoxynucleotide probes that are complementary to specific regions within the bacterial ribosome. The specificity of the probe can be adjusted to fit any taxonomic ranking, from kingdom to subspecies.[32] The generation of probes from genus to species level has been greatly assisted by the access to sequence data and the development of bioinformatic software.[33-36] However, at present the development of strain specific probes is still likely to require sequencing of ribosomal DNA to identify novel sequences.[18]

Ribosomal RNA copies exceed chromosomally encoded genes by approximately 10^4 fold in a single cell and this greatly increases the ability to detect a complementary sequence. Food and faecal samples were analysed using a bifidobacteria genus specific rRNA probe and a hybridisation procedure developed to probe colonies transferred *in situ* onto membrane filters from plates containing semi-selective media.[15] Both isotopic and nonisotopic labelled probes were applied and both were effective, although radioactive probes produced more distinct signals. The technique enumerated between 2.1×10^6 and 2.3×10^7 cfu/g of bifidobacteria from a variety of probiotic dairy products, including cheeses, yogurts, butter and probiotic drinks. While this study did not compare molecular probing results with enumeration based on the use of selective media alone, other studies have made direct comparisons.[37] Faecal samples obtained from adult humans treated with antibiotics were continuously cultured in media containing a prebiotic. The enumeration of bifidobacteria using Bereen's selection media and plate counts concluded that bifidobacteria dominated the culture over a 21 day period. In a parallel experiment within the same study, using a genus-specific rRNA probe on RNA samples isolated from colonies, it was clear that after several days the bifidobacteria were undetectable. The authors concluded that data based on traditional culture techniques may be incomplete and misleading.

Using a fluorochrome labelled rRNA probe, a hybridisation protocol was developed to detect signals within individual cells isolated directly from faeces and transferred to membrane filters.[17] Bifidobacteria counts among ten human subjects ranged from 0.41 to 4.63×10^9 cfu/g of faeces. These data were in excellent agreement with values obtained in a parallel experiment from the same study, in which colonies isolated on a semi-selective media were transferred to glass slides and probed with the same rRNA probe. In samples obtained from the ten subjects, between 0 and 10% of the colonies screened failed to hybridise and were considered to be non-bifidobacteria. In addition, it appeared that all of the bifidobacteria present in faeces could be cultured. In a preliminary study using the same technique to directly enumerate bifidobacteria in the faeces of bottle-fed infants, between 90 and 100% of the faecal bacteria were considered to be bifidobacteria.[18] However, from studies based on traditional enumeration methods using selective media, bifidobacteria were found to represent between 10 and 70% of the total population three days after birth.[38] The authors suggested the discrepancy reflected the higher sensitivity of the molecular method to detect bifidobacteria. The detection of fluorescent probe signals within single cells requires specialist equipment,[39] which may limit the general development of the technique.

A series of 16S rRNA species-specifc probes were developed for the detection of *B. adolescentis, B. bifidum, B. breve, B. infantis* and *B. longum*[40] (US Patent Database). Experiments were performed on pure cultures using radioactive probes. A positive hybridisation was detected with a minimum of 10^3 cells, indicating that sufficient cells exist within an individual colony to allow direct speciation.

2.2.3 Non-ribosomal DNA Probes. A series of species-specific DNA probes were randomly cloned from *B. breve,*[41] *B. longum, B. adolescentis* and *B. animalis.*[42] In the latter study, the probes were used in combination with restriction enzymes to differentiate strains based on distinct RFLP patterns. A genus specific probe generated from whole chromosomal DNA from *B. longum* and *B. adolescentis* was used to enumerate between 1×10^8 and 1×10^{10} bifidobacteria/g of faeces from five human individuals.[14] Cells from colonies were transferred *in situ* onto membrane filters and hybridised to nonisotopic labelled probes. Probe detection used an alkaline phosphatase dye precipitation reaction. The authors compared the method to traditional procedures to enumerate bifidobacteria in which isolates from non-selective media were differentiated using colony and cellular morphology, gram reaction and aerobic and anaerobic growth. Overall, both approaches yielded very similar faecal counts of bifidobacteria but the authors considered the traditional methodology to be cumbersome and time consuming.

2.2.4 Polymerase Chain Reaction and Specific Primers. PCR allows the rapid and specific detection of a wide range of bacterial species.[43] Adavantages of PCR analysis include low costs, high throughput and high sensitivity.

2.2.4.1 Amplification of rDNA sequences. A number of studies have used strain specific PCR primers for the 16S rRNA gene. For example, a *Bifidobacterium* strain (subsequently speciated to *B. lactis*[2]) was identified in the faeces of infants fed the strain in formulated milk powder.[18] No faecal culturing was required as DNA was isolated directly from the faecal microflora prior to PCR. A level of quantification was achieved by determining the detection limit of the PCR assay (10^4 cells/g or 2×10^{-5} % of the total population), using predetermined mixtures of the probiotic with faecel bacteria. The authors compared the sensitivity of the PCR approach to *in situ* hybridisations to DNA sequences, which they considered to be 10^8 cells or 0.1% of the microflora population.

Species and group specific PCR primers were designed for *B. adolescentis, B. angulatum, B. bifidum, B. breve, B. catenulatum/B. pseudocatenulatum* and *B.*

longum/B. infantis and used to identify 43 isolated strains.[44] Primers specific to *B. adolescentis* and *B. longum* were used in PCR reactions with faecal cells without prior DNA purification. This was achieved by the removal of faecal material, which interferes with PCR, through washing and centrifugation.[19] Quantification of the bifidobacteria was attained by determining the maximum sample dilution necessary to achieve a positive PCR result. However, it was evident that the effectiveness of given probes can differ greatly and limit the sensitivity of the technique. For pure cultures of *B. adolescentis* to produce a PCR product after 37 cycles, 10^4 cells were required compared with only two cells for *B. longum*. The two species were among six with low PCR titres from human faeces.

Genus specific PCR primers derived from probes previously applied to *in situ* detection of rRNA[17] were also used to enumerate all bifidobacteria directly from faecel samples.[18] Counts in numerous dairy products were also estimated by testing DNA samples extracted from individual colonies isolated on semi-selective media and grown in broth.[15] However, no enumeration results based on PCR analysis were presented.

An alternative means of strain identification was achieved by restriction enzyme digestion of PCR products formed using 16S rRNA primers.[45] The resulting RFLP patterns were used to track an ingested bifidobacteria strain in several human volunteers. PCR was performed directly on colonies isolated on selective media using microwaves to lyse the cells. The specific PCR product was eluted from the gel, digested with *Hae*II and subjected to a second round of electrophoresis to reveal the RFLP pattern.

2.2.4.2 Random amplified polymorphic DNA (RAPD) PCR. RAPD PCR relies on a single arbitrarily chosen primer to generate a discrete DNA fingerprint of amplified products.[46] PCR using a single primer is a survey of a genome for sites to which it shares homology. The technique is rapid, sensitive and specific as the entire genome of an organism is used as the basis for generating a DNA profile and the variation in size and number of DNA fragments from different strains is the basis for strain differentiation from related species. Although it provides a good level of discrimination between strains and requires no prior knowledge of the genome, there are difficulties in the precise comparison of RAPD PCR fingerprint patterns of different strains from different laboratories. This inter-laboratory variation is primarily due to the use of different primers, employment of different reaction conditions and lack of a standardised approach to the interpretation of the DNA profiles obtained.[4,47]

RAPD PCR was used to characterise fifteen bifidobacteria isolated from rat intestinal tissue.[48] The profiles were compared to those of twenty typed strains representing different *Bifidobacterium* species. The individual RAPD PCR patterns were reproducible and, using a clustering algorithm, a dendogram was constructed to compare the RAPD PCR patterns. All the isolates were differentiated and twelve could be grouped into three basic clusters; however, none showed any relatedness with any of the type strains. The authors suggested that the rat gut isolates belong to bifidobacterial species that have yet to be described.

In another study, RAPD PCR based on five single primer reactions generated highly reproducible profiles for seventy-nine *Bifidobacterium* strains[16]. By combining the profiles obtained for each primer, it was possible to differentiate all the strains tested. Clear clusters were evident for *B. breve, B. bifidum* and *B. adolescentis*. Four other species formed two cluster groups: *B. longum* with *B. infantis* and *B. animalis* with *B. lactis*. In a previous study, PFGE analysis was performed on many of the same strains.[23] In numerous cases, strains with identical PFGE patterns produced different but clustered RAPD PCR profiles. The percentage similarities between strains with identical PFGE profiles ranged from approximately 60 to 85%. The study demonstrated the superior discriminatory power of RAPD PCR compared with PFGE and a possible application in strain speciation. A study undertaken in our laboratory, using two primers in a single PCR reaction, generated unique RAPD PCR profiles for 21 *Bifidobacterium* strains obtained from various culture collections and commercial sources (Table 2). The profiles were then compared with those obtained from five human intestinal bifidobacterial isolates, after construction

Table 2 *Bifidobacteria strains compared using RAPD PCR*

Species	Strain	Source
Bifidobacterium adolescentis	NCFB 2204	Adult intestine
	NCFB 2229	Adult intestine
	NCFB 2230	Adult intestine
	NCFB 2231	Adult intestine
Bifidobacterium angulatum	NCFB 2236	Human faeces
Bifidobacterium bifidum	NCFB 795	Human milk
	NCFB 2203	Infant intestine
Bifidobacterium breve	NCFB 2257	Infant intestine
	NCFB 2258	Infant intestine
	NCIMB 8807	Nursling stools
	NCIMB 8815	Nursling stools
	NCTC 11815	Infant intestine
Bifidobacterium dentium	NCFB 2243	Dental carries
Bifidobacterium infantis	NCFB 2205	Infant intestine
	NCFB 2205	Infant intestine
	NCFB 2256	Infant intestine
Bifidobacterium lactis	Chr. Hansens Bb12	Commercial strain
Bifidobacterium longum	NCFB 2259	Adult intestine
	NCIMB 8809	Nursling stools
	Visby, BB536	Commercial strain
Bifidobacterium pseudocatenulatum	NCIMB 8811	Nursling stools
Bifidobacterium sp.	UCC35612	Adult intestine
	UCC35624	Adult intestine
	UCC35658	Adult intestine
	UCC35687	Adult intestine
	UCC35675	Adult intestine

NCFB: National Collection of Food Bacteria, UK. NCIMB: National Collection of Industrial and Marine Bacteria, UK. NCTC: National Culture Type Collection, UK. UCC: Culture Collection, University College Cork, Ireland. Visby: Laboratorium Wiesby, Germany. Chr. Hansens Laboratory, A/S, Copenhagen, Denmark.

of a dendogram using Gelcompar software (Figure 1). These bifidobacteria were previously isolated from the human gastro-intestinal tract and were obtained from University College Cork (UCC), under a restricted materials transfer agreement. The UCC human isolates generated similar but distinct RAPD PCR profiles that were placed into one cluster group, closest in similarity to *B. longum*. In a previous study using PFGE analysis, the five UCC human isolates generated only two discrete RFLP patterns[26]. These findings support the conclusions of earlier work[16] that RAPD PCR is a more powerful technique compared with PFGE for the discrimination of strains and may have a possible application in strain speciation.

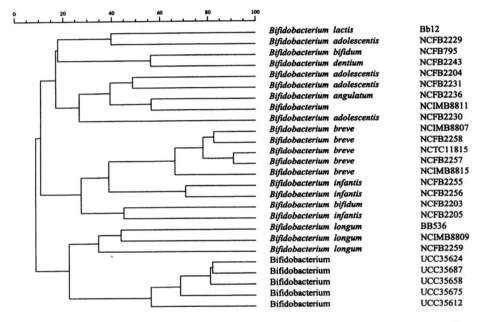

Bifidobacterium lactis	Bb12
Bifidobacterium adolescentis	NCFB2229
Bifidobacterium bifidum	NCFB795
Bifidobacterium dentium	NCFB2243
Bifidobacterium adolescentis	NCFB2204
Bifidobacterium adolescentis	NCFB2231
Bifidobacterium angulatum	NCFB2236
Bifidobacterium	NCIMB8811
Bifidobacterium adolescentis	NCFB2230
Bifidobacterium breve	NCIMB8807
Bifidobacterium breve	NCFB2258
Bifidobacterium breve	NCTC11815
Bifidobacterium breve	NCFB2257
Bifidobacterium breve	NCIMB8815
Bifidobacterium infantis	NCFB2255
Bifidobacterium infantis	NCFB2256
Bifidobacterium bifidum	NCFB2203
Bifidobacterium infantis	NCFB2205
Bifidobacterium longum	BB536
Bifidobacterium longum	NCIMB8809
Bifidobacterium longum	NCFB2259
Bifidobacterium	UCC35624
Bifidobacterium	UCC35687
Bifidobacterium	UCC35658
Bifidobacterium	UCC35675
Bifidobacterium	UCC35612

Figure 1 *Dendogram of RAPD profiles for 26 Bifidobacterium strains, constructed using GelCompar software and grouped using the unweighted pair group algorithm with arithmetic averages (UPGMA)*

3 Conclusion

With increasing interest in and demand for consumer-driven products, much attention has focused on the development of foods containing probiotic microorganisms. A wide range of food products containing bifidobacteria have become commercially available in recent years but it remains to be seen if such products can deliver adequate numbers of viable probiotics to the host. Numerous molecular genetic techniques have been developed to track bifidobacteria in the gastro-intestinal tract through the analysis of faecal microflora. Techniques based on either rRNA hybridisations with labeled probes or specific PCR amplifications have already been used to enumerate bifidobacteria present in dairy food products. However, there is a need to assess the merits of such approaches compared with traditional methods based on colony counts on selective media. Moreover, molecular genetic techniques offer the potential to enumerate bacteria directly without the need to culture. If such approaches prove suitable for routine laboratory analysis, then the task of enumerating bifidobacteria in food products could be greatly simplified.

Acknowledgements

Funding by the European Research and Development Fund and EU grant SMT4-CT98-2235 is gratefully acknowledged. SMcB was supported by a Teagasc Walsh Fellowship.

References

1 S. Naidu, W.R. Bidlack, and R.A. Clemens, *Crit. Rev. Food Sci. Nutr.*, 1999, **38**, 13.
2 J. Prasad, H. Gill, J. Smart and P.K. Gopal, *Int. Dairy J.*, 1998, **8**, 993.
3 W.P. Charteris, P.M. Kelly, L. Morelli and J.K. Collins, *Int. J. Food Microbiol.*, 1997, **35**, 1.
4 D.J. O'Sullivan and M.J. Kullen, *Int. Dairy J.*, 1998, **8**, 513.
5 M.L. Calicchia, C.I.E. Wang, T. Nomura, F. Yotsuzuka and D.W. Osato, *J. Food Prot.*, 1993, **56**, 954.
6 L. Arroyo, L.N. Cotton and J.H. Martin, *Cult. Dairy Prod. J.*, 1995, **30**, 12.
7 K.S. Lim, C.S. Huh, Y.J. Baek and H.U. Kim, *J. Dairy Sci.*, 1995, **78**, 2108.
8 T. McCann, T. Egan and G.H. Weber, *J. Food Prot.*, 1995, **59**, 41.
9 S.C. Ingham, *J. Food Prot.*, 1999, **62**, 77.
10 B. Arany, C.R. Hackney, S.E. Duncan, H. Kator, J. Webster, M. Pierson, J.W. Boling and W.N. Eigel, *J. Food Prot.*, 1995, **58**, 1142.
11 R.I. Dave and N.P. Shah, *J. Dairy Sci.*, 1996, **79**, 1529.
12 B. Pacher and W. Kneifel, *Int. Dairy J.*, 1996, **6**, 43.
13 J.F. Payne, A.E.J. Morris and P. Beers, *J. Appl. Microbiol.*, 1999, **86**, 353.
14 T. Kaneko and H. Kurihara, *J. Dairy Sci.*, 1997, **80**, 1254.
15 P. Kaufmann, A. Pfefferkorn, M. Teuber and L. Meile, *Appl. Environ. Microbiol.*, 1997, **63**, 1268.
16 D. Vincent, D. Roy, F. Mondou and C. Dery, *Int. J. Food Microbiol.*, 1998, **43**, 185.
17 P.A. Langendijk, F. Schut, G.J. Jansen, G.C. Raangs, G.R. Kamphuis, M.H.F. Wilkinson and G.W. Welling, *Appl. Environ. Microbiol.*, 1995, **61**, 3069.
18 R.G. Kok, A.D. Waal, F. Schut, G.W. Welling, G. Weenk and K.J. Hellingwerf, *Appl. Environ. Microbiol.*, 1996, **62**, 3668.
19 R.F.Wang, W.-W. Cao and C.E. Cerniglia, *Appl. Environ. Microbiol.*, 1996, **62**, 1242.
20 C. Stanton, G. Gardiner, P.B. Lynch, J.K. Collins, G. Fitzgerald and R.P. Ross, *Int. Dairy J.*, 1998, **8**, 491.
21 B. Sgorboti, B. Biavati and D. Palenzona, 'The genus *Bifidobacterium*,' Blacky Academic and Professional, London, 1995, p. 279.

22   N. Ishibashi and S. Shimamura, *Food Technol.*, 1993, **47**, 6.

23   D. Roy, P. Ward and G. Champagne, *Int. J. Food Microbiol.*, 1996, **29**, 11.

24   N. Bouget, J.-M. Simonet and B. Descaris, *FEMS Microbiol. Lett.*, 1993, **110**, 11.

25   L. McCartney, W. Wenzhi and G.W. Tannock, *Appl. Environ. Microbiol.*, 1996, **62**, 4608.

26   K. O'Riordan and G.F. Fitzgerald., *FEMS Microbiol. Lett.*, 1997, **156**, 259.

27   M. Farber, *J. Food Protect.*, 1996, **59**, 1091.

28   I. Mangin, N. Bourget, Y. Bouhnik, N. Bisetti, J. Simonet and B. Decaris, *Appl. Environ. Microbiol.*, 1994, **60**, 1451.

29   I. Mangin, N. Bourget and B. Decaris, *Res. Microbiol.*, 1996, **147**, 183.

30   R. Woese, *Microbiol. Rev.*, 1987, **51**, 221.

31   J. Olsen, R. Overbeek, N. Larsen, T.L. Marsh, M.J. McCaughey, M.A. Maciukenas, W.M. Kua, T.J. Macke, Y. Xing and C. Woese, *Nucleic Acids Res.*, 1992, **20**, 2199.

32   R. Frothingham, A.J. Duncan and K.H. Wilson, *Mirobiol. Ecol. Health Dis.*, 1993, **6**, 23.

33   D. Benson, D.J. Lipman and J. Ostell, *Nucleic Acids Res.*, 1993, **21**, 2963.

34   N. Larsen, G.J. Olsen, B.L. Maidak, M.J. McCaughey, R. Overbeek, T.J. Macke, T.L. Marsh and C.R.Woese, *Nucleic Acids Res.*, 1993, **21**, 3021.

35   M. Rice, R. Fuchs, D.G. Higgins, P.J. Stoehr and G.N. Cameron, *Nucleic Acids Res.*, 1993, **21**, 3025.

36   N. Leblond-Bourget, H. Philippe, I. Mangin and B. Decaris, *Int. J. Syst. Bacteriol.*, 1996, **46**, 102.

37   A. Sghir, J.M. Chow and R.I. Mackie, *J. Appl. Microbiol.*, 1998, **85**, 769.

38   M.S. Cooperstock and A.J. Zedd, 'Human Intestinal Flora in Health and Disease', Academic Press, Inc., New York, 1983, p. 79.

39   M.H.F. Wilkinson, G.J. Jansen and D. van der Waaij, *Trends Microbiol.*, 1994, **2**, 485.

40   T. Yamamoto, M. Morotomi and R. Tanaka, *Appl. Environ. Microbiol.*, 1992, **58**, 4076.

41   M. Ito, T. Ohno and R. Tanaka, *Microbiol. Ecol. Health Dis.*, 1992, **5**, 185.

42   I. Mangin, N. Bourget, J.-M. Simonet and B. Decaris, *Res. Microbiol.*, 1995, **146**, 59.

43   J.J. Carrino and H.H. Lee, *J. Microbiol. Methods*, 1995, **23**, 3.

44   T. Matsuki, K. Watanabe, R. Tanaka and H. Oyaizu, *FEMS Microbiol. Lett.*, 1998, **15**, 113.

45   M.J. Kullen, M.M. Amann, M.J. O'Shaughnessy, D.J. O'Sullivan, F.F. Busta and L.J. Brady, *J. Nutr.*, 1997, **127**, 89.

46   J. Welsh and M. McClelland, *Nucleic Acids Res.*, 1990, **18**, 7213.

47   M. Jayaro, B.E. Gillespie and S.P. Oliver, *J. Food Protect.*, 1998, **59**, 615.

48   L. Fanedl, F.V. Nekrep and G. Avgustin, *Can. J. Microbiol.*, 1998, **44**, 1094.

49   Y. Tamine, V.M.E. Marshall and R.K. Robinson, *J. Dairy Res.*, 1995, **62**, 151.

50   D. Knorr, *Trends Food Sci Technol.*, 1998, **9**, 295.

51   P. Dinakar and V.V. Mistry, *J. Dairy Sci.*, 1994, **77**, 2854.

52   M.P. Gomes, F.X. Malcata, F.A.M. Klaver and H.J. Grande, *Neth. Milk Dairy J.*, 1995, **49**, 71.

53   B. Ghoddusi and R.K. Robinson., *Dairy Ind. Int.*, 1996, **61**, 25.

54   M. Gobbetti, A. Corsetti, E. Smacchi, A. Zocchetti, and M. De Angelis, *J. Dairy Sci.*, 1998, **81**, 37.

55   L. Blanchette, D. Roy, G. Belanger, and S.F. Gauthier, *J. Dairy Sci.*, 1996, **79**, 8.

56   D. Roy, I. Mainville and F. Mondou, *Int. Dairy J.*, 1997, **7**, 785.
57   M.P. Gomes and F.X. Malcata, *J. Dairy Sci.*, 1998, **81**, 1492.
58   H. Khalil and E.H. Mansour, *J. Food Sci.*, 1998, **63**, 702.
59   S. Hekmat and D.J. McMahon, *J. Dairy Sci.*, 1992, **75**, 1415.

Lactulose Stimulates Calcium Absorption in Postmenopausal Women Dose-dependently

E.G.H.M. van den Heuvel and E.J. Brink

TNO NUTRITION AND FOOD RESEARCH INSTITUTE,
DEPARTMENT OF PHYSIOLOGY, PO BOX 360, 3700 AJ ZEIST,
THE NETHERLANDS

1 Introduction

Animal studies have indicated that calcium (Ca) absorption is increased by lactulose.[1,2] In an earlier study we found an increase in calcium absorption in adolescents due to the intake of 15g of oligofructose per day.[3] As no results of human studies on the influence of lactulose on Ca absorption were available, a study was conducted in a population group (postmenopausal women) who may benefit from the possible enhancing effect of lactulose on Ca absorption. The aim of this study was to investigate, in postmenopausal healthy women, the possible positive effect of the consumption of lactulose powder on true fractional absorption of Ca in a dose-dependent way.

2 Subject and Methods

2.1 Subjects

Twelve women were selected, who had been postmenopausal for at least 5 years based on a high level of follicle-stimulating hormone and a low level of oestradiol. Their age ranged between 56 and 64 years (mean 60.5). The study protocol was approved by the TNO external Medical Ethics Committee.

2.1.1 Study design and execution of the study. The study was conducted according to the current revision of the Declaration of Helsinki (Somerset West, Republic of South Africa, 1996) and the ICH Guidelines voor Good

Clinical Practice (ICH Topic E6, adopted 01-05-1996 and implemented 17-10-1997). The subjects were instructed to consume the study substance at breakfast during 9 d, on top of their habitual food intake. The study substances consisted of 5 g or 10 g of lactulose powder (Solvay Pharmaceuticals GmbH, Hannover, Germany) or placebo (aspartame) dissolved in 100 ml of water.

The three treatment periods of 9 d each were allocated according to a double-blind randomized cross-over design, separated by 19 d wash-out periods. On the last two days the subjects were housed in the metabolic unit of the institute and the Ca absorption tests were carried out. Therefore, orange juice with 14 mg of ^{44}Ca was drunk immediately after the study substance and just at the start of a standard breakfast with 162 mg of carrier. Within half an hour after the oral administration of ^{44}Ca, 1.2 mg of ^{48}Ca was injected intravenously. From the ICP-MS measurement of the ^{44}Ca/^{43}Ca and ^{48}Ca/^{43}Ca ratios in urine collected before dose administration and over the 36 h after dose aministration, fractional Ca absorption was computed according to the formula reported by Van Dokkum et al.[4]

3 Results

Figure 1 shows the change of Ca absorption (standard error of difference 1.3). Ca absorption was significantly higher during the consumption of 10 g lactulose per day than during the placebo treatment (P<0.01). A significant linear correlation was found between the dose of lactulose and its positive effect on Ca absorption (P<0.01).

Total Ca excretion in 36 h urine did not differ significantly between treatments.

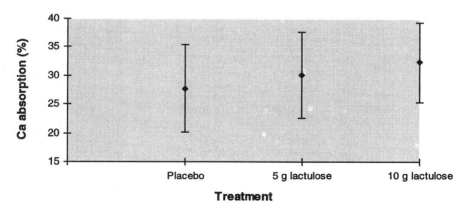

Figure 1 *Changes in Ca absorption (% of intake) in postmenopausal women after a placebo treatment and treatments with 5 g and 10 g of lactulose per day*

4 Conclusion

Lactulose increases Ca absorption in postmenopausal women, without increasing urinary Ca excretion. The positive effect on Ca absorption is significantly related to the dose of lactulose. More research is warranted to explore how lactulose stimulates Ca absorption and whether it is able to improve Ca balance in man and/or to decrease the rate of bone loss with ageing.

References

1   R. Brommage, C. Binacua, S. Antille and A.L. Carrie, *J. Nutr.*, 1993, **123**, 2186.
2   A.M. Heijnen, E.J. Brink, A.G. Lemmens and A.C. Beynen, *Br. J. Nutr.*, 1993, **70**, 747.
3   E.G.H.M. van den Heuvel, Th. Muijs, W. van Dokkum and G. Schaafsma, *Am. J. Clin. Nutr.*, 1999, **69**, 544.
4   W. van Dokkum, V. de La Gueronniere, G. Schaafsma, C. Bouley, J. Luten and C. Latge, *Br. J. Nutr.*, 1996, **75**, 893.

3 Phytochemicals

Plants as Functional Foods

A.F. Walker

THE HUGH SINCLAIR UNIT OF HUMAN NUTRITION,
DEPARTMENT OF FOOD SCIENCE & TECHNOLOGY,
THE UNIVERSITY OF READING RG6 6AP, UK

1 Introduction

Over the last decade there has been a quiet revolution in thinking amongst nutritional scientists. This has culminated in equal research attention being placed on optimal nutrition as on nutrition to avoid deficiency disease (essentiality). The realisation that avoidance of chronic disease and protection against environmental toxins is as much part of the subject of human nutrition as is the prevention of diseases of undernutrition, such as scurvy, has re-adjusted the place of human nutrition in modern medicine, as well as increasing the responsibility of the food industry for the health of the population.

In striving to understand the elements of an optimal diet for health, attention was first focused on the macronutrients, in particular the fat content of the diet. More recently, research into optimal intake of micronutrients for health has gathered momentum. These studies were initially directed at nutrient protection against free-radical cellular damage. Indeed, studies of dietary nutrient antioxidants have led the pioneering work in optimal nutrition. In only a single decade we have seen enormous expansion of research interest in optimal nutrition, ranging from the merits of enhanced intakes of vitamins, to the value of dietary non-essential nutrients of plant origin, called phytochemicals. These phytochemicals are represented in nature by various groups of structures comprising a myriad of individual compounds.

Being both a nutritional scientist and a qualified practitioner of phyto-therapy (herbal medicine), I am in a position to comment on phytochemicals in optimal nutrition and the interface between these two subjects. In this paper, the role of phytochemicals in optimal nutrition, including their antioxidant activity, is discussed. In particular, the ways in which phytochemicals contribute to human health, in a preventative and therapeutic sense, are highlighted.

Plant materials consumed by humans provide a range of bio-activity. They include plants that provide energy for human sustenance, plants used as fruit and vegetables that provide nutrients and phytochemicals in varying amounts, plants used as culinary herbs and spices and those used as herbal medicines. The multi-faceted physiological actions of plant foods and medicines reflect the complex bioactive constituent profiles present in all plant materials. The traditional procedures used for concentration of phytochemicals from these materials for therapeutic purposes include aqueous and aqueous-alcoholic solvent extraction. However, effective phytotherapy requires consumption of adequate doses, consumed on a daily basis for sufficient duration. Extracts of foods and tonic medicinal plants have potential for phytonutrient enrichment of functional foods. However, such enhancement poses a challenge to the food industry to produce palatable functional food products.

2 Changing Concepts of Human Nutrition

A functional food may be defined as a food with an enhanced physiological action beyond that expected of a normal food. The reason why many food companies are interested in developing functional foods directly relates to the changing concepts of human nutrition over the last decade. Earlier nutritional concepts were confined to the concept of essentiality. That is, the avoidance of deficiency diseases, using target dietary intakes for the various human groups such as the RNIs (Reference Nutrient Intakes).[1] With the increasing evidence that enhanced nutrition is linked to the prevention of chronic disease, the concept of optimal nutrition became established. Arguably this new concept was initially launched in the 1970s, with the finding of links between intake of fat and heart disease through Keys' Seven Countries Study.[2] Hence the early research in optimal nutrition was focused on the macronutrients, particularly the balance of the main energy providers; fat, protein, carbohydrate. From this research, developed the dietary guidelines for healthy eating from the 1970s onwards. It was, however, the antioxidant nutrients which provided the impetus for the development of optimal nutrition of micronutrients.

In the 1980s, the concept of nutritional optimality was fuelled by the realisation that free radicals were produced by metabolism (especially that involving oxygen) within the body, as well as in food stored badly in an atmosphere of oxygen. The many experiments, which have been conducted since the late 1980s, have substantiated the importance of dietary antioxidants in reducing the adverse impact of toxic free radicals on health, and have fuelled further research into optimal nutrition.[3] Studies on nutrients were soon augmented by investigations into the accessory factors to vitamins, which had long been purported, but not proven, to play a supportive role in vitamin activity. These accessory factors are the phytochemicals such as the flavonoids. A decade further on and phytochemicals are now fully established as part of the body's antioxidants armamentaria for minimising toxic free radical activity.

Fruit and vegetables comprise the food group which epidemiological studies repeatedly show to reduce the risk of chronic disease, when consumed in adequate quantities.[4] Studies in the 1970s of the health protective effects of fruits and vegetables concentrated on their contents of dietary fibre. It is now clear that dietary fibre has a important role to play in human nutrition, if less so than researchers predicted from their first findings. However, it has now emerged that the health benefits of fruit and vegetables are largely due to the minor components of this food group, in particular the antioxidant vitamins, but strongly supported by the large numbers of phytochemicals, some with even greater antioxidant properties than the antioxidant vitamins.

The first group of phytochemicals which received attention was the carotenoids. These are a highly antioxidant group but, nevertheless, their presence in foods does not account for the full health protective effects of fruit and vegetables. More recent interest has turned to other non-essential nutrients and, in particular, the group known as flavonoids.[5] This group may account for a large proportion of the protective and therapeutic effects of plant materials. Examples of the physiological activity of fruit and vegetables can be seen in Table 1. Not only can their physiological effects help to prevent disease, some can be used in therapy. For each of the fruits and vegetables shown in Table 1, preliminary studies indicate efficacy in treatment of the disorders shown.

Table 1   *Plant foods with preventative/therapeutic effects in human studies*

Plant food	Condition	Reference
Artichokes	Hypercholesterolaemia	6
Bilberries	Retinopathy	7
Broccoli	Cancer prevention	8
Cabbage	Peptic ulcers	9
Cherries	Gout	10
Cranberries	Cystitis	11

All plant extracts are antioxidant to a degree, since botanical processes are mostly conducted in a reducing internal environment conducive to optimal function of plant cells. However, plant tissues from some species are more antioxidant than others. Some examples taken from data of Pietta *et al.*[12] are shown in Figure 1. These include Bilberries (*Vaccinium myrtillus*), the beverage Green tea (*Camilla sinensis*) and some herbal medicines. It is interesting to note that Bilberries and Green tea have antioxidant activity greatly in excess of some medicinal plants; hence, Bilberries are particularly antioxidant compared with the medicinal herbs Hawthorn (Crataegus spp.), Chinese ginseng (*Panax ginseng*) and Red clover *(Trifolium pratense)*. Since these herbs have proven therapeutic effects, mechanisms other than antioxidant activity must contribute to their activity.

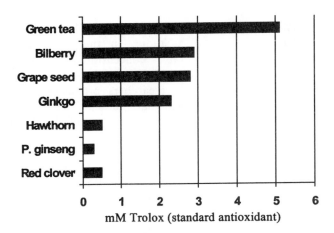

Figure 1 *Total antioxidant activity of plant extracts* (0.5 mg/ml)[12]

3 The Basis of Phytotherapy

The basis of Phytotherapy as used in western herbal medicine is the promotion of homeostasis.[13] Homeostasis can be defined as the maintenance of the composition of the internal environment of the body fluids within narrow limits, consistent with the optimal function of the body's cells. When applying phytotherapy for disease conditions, it is important to consider the influences on the group of cells which are showing dysfunction or pathology. Cell dysfunction occurs before pathology and it is at this stage that the phytotherapist would prefer to intervene, prior to pathology. Cell dysfunction may be due to a number of influences (stressors). Obviously, a cell may be stressed if it lacks any one of the forty or more nutrients which are required for its function. In addition, cell function may be optimised by the presence of nutrient accessory factors – the phytochemicals, which are interchangeable to a degree. Many phytochemicals such as the flavonoids are highly antioxidant and support the action of the antioxidant vitamins. Although we are only just beginning to understand the mechanisms by which phytochemicals exerts health benefits apart from their antioxidant properties, there is increasing evidence of receptor activity among them: *e.g.* the isoflavonoids bind to oestrogen receptors.[14] Receptor activity may account for the clinical observations that low doses of phytochemicals may have profound effects taken over a period of time, such as lessening adverse menopausal symptoms.[15]

Other stressing influences on the cell might be a lack of oxygen, which stimulates the cell to produce eicosanoids, which trigger responses in surrounding tissue to promote blood flow and hence to ameliorate the situation. Stress on the cell will then only continue if the 'call' for enhanced blood flow goes unheeded. This might occur, for example, in peripheral vascular disease, where partial arterial occlusion due to atherosclerosis hinders blood flow. Another stressor on the cell might be the diminished capacity of the body to remove metabolites (toxins) at a fast enough rate from a particular site. Of course, accumulation of toxins may result from environmental exposure to toxins, but in most disease conditions, the toxic build up from metabolism is most important. Toxins may build-up particularly in regions of the body that are poorly vascularated, such as the joints.

Cells of the body may also be subject to specific stressors. For example, breast cells may be especially susceptible to high concentrations of circulating oestrogen. Certainly, it is well known that the risk of breast cancer is enhanced against a background of high oestrogenicity (free oestrogen, with low sex hormone binding globulin).[4] Another cell stressor is the pituitary hormone prolactin, which is itself a stress response hormone, being markedly elevated in the blood soon after a stressful event which provokes the anxiety response. Prolonged stress of long duration will lead to continuously high plasma prolactin concentration. Raised prolactin acts as a cell stressor in hormone-dependent tissues. These include breast and prostate cells. These, along with other mechanisms, are thought to enhance risk of prostate and breast cancer. In the case of breast cancer, we have evidence from animal studies and epidemiology that high intake of Cruciferous vegetables such as broccoli may be helpful in reducing the risk of breast cancer.[8] The mechanism has been suggested to be related to the way in which broccoli phytochemicals, including indoles, induce phase II liver enzymes, so enhancing detoxication and promoting oestrogen disposal.[6]

Key homeostatic systems of importance in phytotherapy include the digestive system. Digestion is important in the provision of nutrients, but also plays a role in reducing allergen load through adequately functioning digestive glands and gastrointestinal tract co-ordination. Poor digestive performance will lead to large molecules being presented to the lymphoid tissue in the ileum at the location of the Peyer's Patches. Indeed, the traditional saying of herbalists that 'if the gut is not right, the rest of the body is not right' is being borne out by modern studies including those on pro- and prebiotics which is dealt with elsewhere in this Symposium. Circulation is another important aspect of homeostasis. It is this system which transports the nutritive and toxic molecules between organs and cells. If tissues suffer poor blood supply then metabolite accumulation might be anticipated, leading to cell stress.

The main responsibility for elimination of toxins in the body lies with two organs, the liver and kidneys. The liver is responsible for bio-transformation of toxins and bio-active compounds entering the body from the digestive tract, but it also deals with the disposal of the end products of metabolism, *e.g.* insulin and oestrogen disposal. The detoxication function of the liver reduces molecular mass (phase 1 reactions) and enhances water solubility by conjugation (phase II reactions). These processes facilitate the removal of metabolites and toxins from the body *via* the kidneys. Larger molecules such as steroids, which are too big to be excreted via the kidneys, are passed into the gut via the bile and are eliminated in the faeces. However, some of these large molecules suffer at least a degree of recycling (the enterohepatic circulation), the extent depending on the composition of the diet. For example, a diet high in dietary fibre will bind certain high molecular weight compounds such as oestrogen and carry them out of the body in the faeces, hence reducing recycling.[17] Thus it is that diets high in dietary fibre reduce circulating oestrogen levels. Therefore, elimination via the gut is also important, although to a lesser degree than that via liver and kidneys. Finally, skin elimination via sweat should also be taken into account. It is notable that, in the clinical situation, patients suffering from skin disease, especially eczema, often report an inability to sweat freely. In phytotherapy, sweating may be promoted by the use of herbs with diaphoretic properties, such as Yarrow (*Achillea millefolium*) and Limeflowers (*Tilia europea*).

Regulatory and control mechanisms are very important in homeostatic control, and these involve receptor mechanisms which can be easily targeted with small amounts of phytochemicals present in a typical daily dosage of plant extract. Of particular interest are the nervous and endocrine systems and there are good examples of herbal extracts which moderate them. For example, it is now well documented that isoflavonoid phytoestrogens from soya can bind to oestrogen receptors.[18] Under conditions of excessive oestrogenicity (effective oestrogen), which manifests in symptoms such as breast pain, phytoestrogens compete with circulating oestrogen and reduce its effectiveness and hence alleviate the symptoms. Hence, isoflavonoids can act in an anti-oestrogen capacity. However, of more interest to women at the menopausal stage of life is their oestrogenic activity. By occupying the oestrogen receptor on the cell wall of target tissues, adverse effects of low circulating oestrogen, such as hot flushes, are minimised.[15]

4 Medicine as Food, Food as Medicine

Plants provide mankind with a wide spectrum of nutritive, health-promoting and therapeutic materials. At one end of the spectrum, we have foods which provide energy and fibre sources, which may be low in phytochemicals. Fruit and vegetables provide a wide spectrum of phytochemicals, but have been selected for palatability, crispness and lack of bitterness: attributes which tend to lower their phytochemical content. Culinary herbs and spices, however, are rich sources of phytochemicals and, as such, are markedly pungent, aromatic, bitter or otherwise strong tasting. Similarly rich sources of phytochemicals are found among the tonic herbal medicines, which are not used for culinary purposes. The Materia Medica of western herbal medicine is largely composed of 'tonic' herbal medicines, and includes some fruit and vegetables and all of the culinary herbs and spices, as well as herbs not used for culinary purposes. Tonic herbal medicines contain active principles, some of which are common to fruit and vegetables and culinary herbs and spices. Therefore, the physiological actions of these tonic herbs have more in common with nutrients than they have to high-dose single active chemicals used in orthodox medicine (the so-called 'magic bullet'). The application of phytotherapy through the use of tonic herbs, results in benefits to all cells of the body, by the provision of non-essential nutrients (accessory factors) to support the action of vitamins and minerals. Most herbal medicines used on a daily basis by phytotherapists are in the tonic herbal medicine category and might be considered as candidate plant extracts for inclusion in functional foods. A group of medicines with potent action, due mainly to their contents of alkaloids, are in a separate category, and these are restricted in their use by qualified herbalists in the UK under the 1968 Medicines Act (see Licensing of herbal remedies, in Bartram).[19] This group, which includes the two tropane-alkaloid containing herbs, Deadly nightshade (*Atropa belladonna*) and Henbane (*Hyoscyamus nigra*), is totally unsuitable for inclusion in functional foods.

Finally, at the end of the spectrum, we have toxic plant materials, whose physiological actions give rise to adverse effects through highly bio-active material. These herbs are recognised as toxic in all cultures and consumption is avoided through education. In the UK, certain potent herbal medicines are placed in this category, for example Foxglove (*Digitalis purpuria*), which is an accumulative poison for the heart and Opium poppy (*Papaver somniferum*), which is an addictive opiate. In the UK, such toxic plant materials are banned from use by herbalists under the 1968 Medicines Act.

Table 2 shows evidence of the low acute toxicity of the essential oils from some selected foods and herbal medicines. For many plants or plant extracts, LD50s (dose to cause death of 50% of the animals tested) can be found in the literature, but such data is not gathered systematically. The data comes from the efforts of individual academics working often in isolation, and is gathered in an entirely haphazard fashion. Table 2 was compiled from Bradley[20] and ESCOP[21] (European Scientific Cooperative on Phytotherapy) Monographs. The herbs are ranked according to their food use, followed by beverage and culinary use, and end with those solely used as medicinal plants. Table 2 shows that the LD50s of plant extracts (gathered at random from the two sources indicated) are large, indicating low toxicity of essential oils, which are very concentrated sources of phytochemicals. An interesting point to note in Table 2 is that the acute toxicity of essential oils from foods such as Ginger (*Zingiber officinale*) and culinary herbs such as Thyme (*Thymus vulgaris*) are similar. Indeed, Valerian (*Valeriana officinalis*) which is only used as a herbal medicine and has no culinary application whatsoever, has the largest LD50 in Table 2, indicating its very low toxicity (the larger the LD50, the lower the toxicity).

All herbs in Table 2 are used on a day-to-day basis by practising medical herbalists for treatment of their patients. Several monographs have been published on each herb, including German Commission E monographs,[22] which document their indications. Some examples of the indications for use of these herbs can be seen in Table 2: for each plant only one action and one indication is given in Table 2. Practitioners of herbal medicine understand each herb to have a profile of actions and indications, involving several organ systems. This multiple action reflects the complex composition of each herb, being composed of many phytochemical entities, each of which may have more than one action. Hence, it should not be surprising to find that the physiological action of the entire plant extract (its therapeutic profile) is broad. This point is often puzzling to orthodox practitioners, accustomed to using a single chemical drug with a well-defined mechanism of action. This point will be expanded further in the next Section.

Table 2   *Some LD 50s for essential oils with actions and indications*

Plant	Used as	LD50 (g/kg body weight	Action	Indication
Ginger	F B C M	>5	Antiemetic	Nausea
Fennel	B C M	4.5	Carminative	Infant colic
Chamomile	B M	2.5	Antiinflammatory	Childhood irritability
Aniseed	C M	2.7	Expectorant	Cough
Thyme	C M	2.8	Antiseptic	Bronchitis
Valerian	M	15	Sedative	Anxiety

B, beverage; C, culinary herb/spice; F, food; M, herbal medicine
Compiled from: Bradley,[20] ESCOP[21] and Chevallier.[23]

Table 3 shows the LD50 for dried plant materials, which, in contrast to the essential oils, are much closer to the herbal extracts of the entire plant, normally used in phytotherapy. Again, Table 3 is ordered according to food, beverage and culinary use. Hence, Nettle leaf (*Urtica dioica*) is used as a food, beverage and medicine, German chamomile (*Matricaria recutita*) is used as a beverage and medicine and Devil's claw (*Harpagophytum procumbens*) is only used as a medicine. Again, the LD50s show low acute toxicity whether the use is as a food or a medicine. Indeed, two well-known tonic herbs, Siberian ginseng (*Eleutherococcus senticosus*) and Devil's claw have LD50s in excess of 12 g per kilogram of body weight. Again, all the herbs in Table 3, apart from Blackcurrant leaf (*Ribes nigrum*), are used on a daily basis by phytotherapists in the UK. Blackcurrant leaf is commonly used as an anti-inflammatory and diuretic agent in Germany for arthritis.[22] One key action and indication only of these plants is shown in Table 3. Again, each plant contains a profile of active constituents, each with physiological activity and, therefore, the crude plant extract would be expected to have a broad therapeutic profile.

Table 3   *Some LD50s of dried plant materials with actions and indications*

Plant	Used as	LD50 (g/kg body weight	Action	Indication
Nettle leaf	F B M	>2	Alterative	Eczema
Dandelion leaf	F M	3–6	Diuretic	Fluid retention
Chamomile	B M	>5	Anti-inflammatory	Gastritis
Hawthorn	F M	6	Hypotensive	Hypertension
Siberian ginseng	B M	25	Tonic	Stress
Blackcurrant leaf	M	>3	Diuretic	Gout
Devil's claw	M	13.5	Anti-inflammatory	Arthritis

B, beverage; C, culinary herb/spice; F, food; M, herbal medicine
Compiled from: Bradley,[20] ESCOP[21] and Chevallier. [23]

5 Bioactivity of Phytochemicals

The wide array of bioactive phytochemicals found in plants can be divided into a number of chemical groupings, examples of which are shown in Table 4. Most of these groups are equally represented in plant foods and plant medicines. For example, salicylates with anti-inflammatory properties are found in Strawberries and in Meadowsweet (*Filipendula ulmaria*), a herb used extensively by phytotherapists for treatment of arthritis. Diosgenin, a steroidal saponin with progesterone-like and anti-inflammatory action, can be obtained from Fenugreek (*Trigonella foenum-graecum*), a culinary spice, as well as from the Wild yam (*Dioscorea villosa*) used as medicine. Diosgenin from the Wild yam was the original source of oestrogen used in the contraceptive pill, having been subjected to a chemical modification.

Table 4 *Some bioactive phytochemicals found in plants*

Chemical group	Example	Action
Aromatic acids	Salicylates	Anti-inflammatory
Anthraquinones	Sennosides	Laxative
Triterponoid saponins	Glycyrrhizin	Vulnerary
Steroidal saponins	Diosgenin	Anti-inflammatory
Alkaloids	Caffeine	CNS stimulant
Flavonoids	Genistein	Oestrogenic

The flavonoids are a large phytochemical family of several hundreds of compounds. Quercetin found in onions, is also found in many non-food plants. Other groups of flavonoids found in common in fruit and vegetables and medicinal plants are shown in Table 5. The specific examples given have all been shown to have physiological or pharmacological effects. Some are widely distributed throughout the plant kingdom: for example, apigenin is found in celery, apples and also the herbal medicines Chamomile and St John's wort (*Hypericum perforatum*).

Genistein shown in Table 4 is an isoflavonoid with oestrogenic properties, found in soyabeans. Many studies now confirm the oestrogenic effects of isoflavonoids, also found in several herbal remedies used traditionally for women's health problems: Red clover (*Trifolium pratense*), Lucerne (*Medicago sativa*), Black cohosh (*Cimicifuga racemosa*).[23] The properties of the isoflavonoids are currently the subject of considerable research interest and will be dealt with in depth later in this Symposium. The ability of isoflavonoids to normalise homeostasis according to the state of the body as mentioned earlier (the anti-oestrogenic effect at high circulating oestrogen levels and pro-oestrogenic effect at low oestrogen levels) is an action termed 'adaptogenic' by herbalists.

Table 5   Some flavonoids found in fruit and vegetables

Flavonoid group	Example	Food
Flavanone	Naringin	Citrus fruits
Flavone	Apigenin	Celery
Flavonol	Quercetin	Onions
Anthocyanins	Cyanidin	Grapes
Phenyl propanoids	Ferulic acid	Spinach

Table 6 lists some of the documented actions of flavonoids. Most of these actions have been demonstrated in animal models and not in clinical studies, that is, until recently. The Zutphen Elderly Study was a prospective epidemiological study which was conducted over a five-year period on 805 elderly men, who had previous heart attacks.[24] In this study, a 68% reduction in mortality was noted as a significant outcome for those men on high intakes of flavonoids versus those on low intakes (the men were divided into quintiles according to flavonoid intake). Foods contributing most to flavonoid intake were black tea, onions and apples. Piecing together the in vitro, animal and human studies, there is little doubt that flavonoids contribute greatly to the overall protective effect of fruit and vegetables for many chronic diseases.

Table 6   Some documented actions of flavonoids

Antioxidant	Oestrogenic
Anti-allergenic	Diuretic
Anti-spasmodic	Hypotensive
Anti-inflammatory	Membrane stabilisation
Anti-neoplastic	Antiseptic
Cardiac/circulatory stimulant	

6 Multiple Actives: Multifaceted Action

Each plant contains a wide profile of constituents, many of which have physiological activity. Two examples of plants used in phytotherapy will illustrate this point. The first is Hawthorn (Crataegus spp.), which is used in treatment of cardio-vascular disorders, principally for its flavonoid content. Food use of Hawthorn berries is documented and the leaves have been used in the past as a beverage (tea). Table 7 shows some of the active constituents of Hawthorn. It is the flavonoid content of Hawthorn which is considered to be its most biologically active phytochemical group. Hence, extracts of the herb are standardised on one of the characteristic flavonoids in this species, namely Vitexin. Flavonoids are purported to contribute to the hypotensive effect of Hawthorn.[25]

Table 7 *Some active constituents of Hawthorn (Crataegus spp.)*[26]

Chemical group	Specific phytochemicals
Amines	Phenylethylamine, methoxyphenethylamine, tyramine
Flavonoids	Flavonols: kaempferol, quercetin
	Flavones: apigenin, luteolin, rutin hyperoside,
	vitexin glycosides, orientin glycosides, procyanidins
Tannins	Proanthocyanidins
Miscellaneous	Cyanogenic glycosides, saponins

At the University of Reading, we conducted a pilot study on the effect on resting diastolic blood pressure of a daily supplement of 600 mg of magnesium or 500 mg of standardised extract of Hawthorn administered to 38 mildly hypertensive subjects.[27] This double-blind placebo-controlled parallel study showed, by ANOVA factorial contrast analysis, that Hawthorn was effective in lowering blood pressure at less than 10% level of significance (p=0.081) (Figure 2). Owing to the small size of the groups involved and the difference in the blood pressures at baseline, we were not able to achieve significance at the planned 5% level. However, the trend shown by the data warrants follow-up with a full clinical study into the hypotensive effects of Hawthorn.

Traditionally, Hawthorn is used for treatment of cardiovascular problems: to reduce atherosclerosis, to lower blood pressure, to strengthen myocardial function and to normalise heart rhythm. It is regarded by phytotherapists as particularly useful for improving blood circulation to the legs in intermittent claudication. The narrow focus of the action of Hawthorn on heart and circulation is unusual for a tonic herbal medicine.

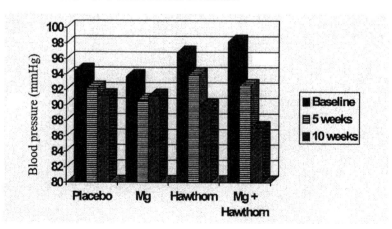

Figure 2 *A pilot study of the effect on resting diastolic blood pressure of a daily supplement of 600 mg magnesium and/or 500 mg standardised extract of Hawthorn (Crataegus spp) on 38 mildly hypertensive subjects: a double-blind, placebo-controlled parallel study*[27]

It is much more common to find that tonic plant materials have a broad spectrum of physiological effects. This is certainly the case for St John's wort, the best researched of all the tonic herbal medicines, which has been used traditionally to benefit the health of all the body systems, even though its primary focus is the nervous system.[23] Plant extracts with broad therapeutic profiles should not come as a surprise when one considers that phytochemicals act more like nutrients than like modern drugs. Hence, their consumption should benefit all cells of the body, as they act in a nutrient role as accessory factors to the essential nutrients. Indeed, the term 'herbal nutrition' better describes the action of tonic herbs, rather than herbal medicine.

Garlic is a good example of a plant composed of multiple active constituents, showing multi-faceted action when ingested. Table 8 shows some of the sulfur compounds found in garlic and Table 9 some of the physiological actions of the herb. Although the action of Garlic is focused on the cardiovascular system, this focus is multi-faceted; hence we see that Garlic has hypocholesterolaemic, hypotensive, anti-thrombotic and anti-oxidant properties. However, as well as that, Garlic extract is strongly antimicrobial. In fact, in traditional herbal medicine, the anti-microbial properties of Garlic were considered as the focus of therapeutic use. The extract has long been known as effective against viruses, bacteria and fungi. Hence, various infections are treated with Garlic, and particularly those of the lungs, as the sulfur compounds are volatile and will pass out of the body on the breath. It is only with modern insight and the results of scientific research, that phytotherapists have come to realise the potential of garlic for the cardiovascular system.

Table 8   *Some sulfur compounds in Garlic*[28]

3,5-Diethyl-1,2,4-trithiolane	Diallyl sulfide
Allyl 1-propenyl disulfide	Diallyl trisulfide
Allyl 1-propenyl trisulfide	Methyl allyl trisulfide
Allyl alcohol	Methyl allyl disulfide
Diallyl disulfide	*S*-allylcysteine

Table 9   *Physiological actions of Garlic*[28]

Hypocholesterolaemic
Hypotensive
Antithrombotic (fibrinolytic, inhibits platelet aggregation)
Antioxidant (reduces LDL oxidation)
Anticarcinogenic/ immune enhancing
Antimicrobial (viruses, bacteria, fungi)
Antihelmintic
Anti-inflammatory, expectorant, carminative

7 Using Plants for Physiological Effect

Before going on to consider the possibilities of developing functional foods using plant extracts, it is important to consider the ways in which plants are used currently in therapy. In the main, herbalists from all cultural traditions use extracts of plants rather than whole plant tissues for treatment of their patients. Foods may provide therapeutic potential, but the limiting factor in their therapeutic use is regular consumption. Fresh or dried plant material may be used as an infusion or decoction. A decoction is made by boiling the plant material with water: a process used for woody materials such as barks and roots. Juices can be made from fruit and vegetables, but also there are juices available from some medicinal herbs. Most of these are manufactured in Germany. The most popular preparations used by practising medical herbalists are aqueous-alcoholic extracts. These extracts are made according to the British Herbal Pharmacopeia[29] and termed tinctures. These are typically a 1:5 extract: 1 part of dried herb by weight to 5 parts of aqueous alcohol by volume. The maceration of the plant material with the solvent is carried out at ambient conditions over at least a two-week period. Aqueous-alcohol is an ideal extraction and preservation medium. It has the advantage over other methods of phytochemical extraction in that no heat treatment is involved, thus minimising changes in phytochemical structural conjugation. It is also interesting to note that both traditional tincture-making procedures and modern chromatographic analysis employ aqueous-alcoholic solvent systems to optimise extraction of phytochemicals from plant materials.

An increasingly popular form of plant extract used in phytotherapy is the glycerol extract. However, the advantages and disadvantages of glycerol extraction, a highly hydrophilic solvent, have not been studied systematically. Glycerol imparts a sweet taste to a bitter herbal medicine, thereby increasing its palatability. Herbal tablets may be prepared from powdered, dried plant material and therefore have low potency. High-potency preparations are made from dried herbal extracts: the extract being prepared in the same way as a tincture, which is then dried. The best quality herbal extracts are standardised to contain a minimum quantity of one or two key active constituents of the plant. These preparations are much more expensive than powdered herb tablets, but lower doses are necessary.

Typical adult doses of plant materials used in phytotherapy may be found in the British Herbal Pharmacopoeia.[29] On a daily basis, dosage for an adult might be 9 g of dried herb as infusion or decoction, or 15 ml of herbal tincture, which is equivalent to 3 g of dried herb. The higher dose of dried herb consumed after water-extraction is likely to reflect the greater extraction rate of phytochemicals from plant material by aqueous-alcohol. In using plants for their therapeutic effect, it is important that they are used regularly, on a daily basis. The patient also needs to be informed that health benefits are not likely to be immediate. Generally, the use of plants as therapeutic agents requires a duration of about six weeks of treatment before benefit is noted. Of course, there are exceptions, such as in the treatment of gastritis, where direct application of the plant extract to the stomach wall leads to a rapid response. Conversely, conditions such as long-standing osteoarthritis may not respond for three months or more.

My outline of the traditional use of herbal extracts in treatment of disease, highlights some general principles of phytotherapy common to all cultural traditions. These principles need to be borne in mind if developing functional foods. The title of my talk 'Plants as Functional Foods' could have been interpreted as applying just to fruit and vegetables, as these clearly have proven physiological activity, with health benefits and some are even used in therapy. However, as a functional food is defined as a food providing a health benefit beyond that expected of a normal food, I have interpreted the title to mean enrichment of foods with plant extracts. If plant extracts are to be used to enrich foods with the aim of providing the consumer with greater choice of foods for health maintenance, then the following points will need to be addressed.

1. The physiological action of a plant extract may take weeks of consumption to manifest.
2. Consumption should be on a regular (usually daily) basis.
3. Strong flavours and odours are characteristic of phytochemical extracts: these need to be overcome in food product development.
4. Colour problems: most plant extracts contain phenolics which oxidise to form dark brown polymeric compounds when exposed to oxygen.

8 Conclusion

Over the last decade, the subject of human nutrition has extended from its original base of ensuring adequate nutrition for prevention of deficiency diseases to embrace optimal nutrition. This movement was driven by the realisation through scientific investigation, of the importance of antioxidants to human health. Antioxidants count for a large part of the protective and therapeutic value of fruits and vegetables, whose merits are now recognised in dietary guidelines throughout the world. The value of fruits and vegetables goes beyond their sum of antioxidant nutrients, to include the non-essential nutrients, phytochemicals. Phytochemicals are now fully recognised as comprising a wide range of antioxidant substances, many of which support the antioxidant activities of the vitamins C and E in their preventative roles.

To use plants as functional foods implies the development of food products having an enhanced level of phytochemicals over and above that present in normal fruit and vegetables. Bio-active minor components of plants can be extracted using traditional procedures such as aqueous-alcoholic maceration. These extracts have the potential for enriching foods to make functional foods. However, the challenge to the food industry will be to produce palatable products with acceptable sensory characteristics that deliver an effective dose, for consumption as a regular part of the diet.

References

1 DH (Department of Health), 'Dietary Reference Values for Food Energy and Nutrients for the United Kingdom', Report on Health and Social Subjects No. 41, HMSO, London, 1991.

2 A. Keys, *Circulation*, 1970, **41**, 1.

3 A. T. Diplock, *Br. J. Nutr.* 1998, **80**, Suppl. 1, S77.

4 DH (Department of Health), 'Nutritional Aspects of the Development of Cancer', Report on Health and Social Subject No. 48, HMSO, London, 1998.

5 M. G. L. Hertog, P. C. H. Hollman, B. van de Putte, *J. Agric. Food Chem.*, 1993, **41**, 1242.

6 A. Lietti, *Fitoterapia*, 1977, **48**, 153.

7 A. Scharrer, M. Ober, *Klin. Monatsbl. Augenheikd.*, 1981, **178**, 386.

8 F. Levi, C. La Vecchia, C. Gulie, E. Negri, *Nutr. Cancer*, 1993, **19**, 327.

9 G. Cheney, *Calif. Med.*, 1949, **70**, 10.

10 L. W. Blau, *Texas Rep. Biol. Med.*,1950, **8**, 309.

11 J. Avorn, M. Monane, J. H. Gurwitz *et al.*, *J.A.M.A.*, 1994, **272**, 751.

12 P. Pietta, P. Simonetti, P. Mauri, *J. Agric. Food Chem.*, 1998, **46**, 4487.

13 D. Hoffmann, 'The Holistic Herbal', 2nd edition, Element Books Ltd, Longmead, Shaftesbury, Dorset, 1986.

14 D. C. Knight, J. A. Eden, *Obstet. Gynecol.*, 1996, **87**, 897.

15 A. L. Murkies *et al.*, *Maturitas*, 1995, **21**, 189.

16 J. J. Michnovicz, H. L. Bradlow, *J.N.C.I.*, 1990, **82**, 947.

17 B. R. Goldin, S. L. Gorbach, *Adv. Exp. Med. Biol.*, 1994, **364**, 35.

18 V. C. Jordan, *Mol. Cell Endrocrinol.*,1990, **74**, C91.

19 T. Bartram, 'Encyclopedia of Herbal Medicine', Grace Publishers, Mulberry Court, Stour Rd, Christchurch, Dorset BH23 1PS, UK, 1995.

20 P. R. Bradley, 'British Herbal Compendium, Volume 1, A Handbook of Scientific Information on Widely Used Plant Drugs'. British Herbal Medicine Association, PO Box 304, Bournemouth, Dorset BH7 6JZ, UK, 1992.

21 ESCOP (European Scientific Cooperative on Phytotherapy), 'Monographs on the Medicinal Uses of Plant Drugs', ESCOP Secretariat, Argyle House, Gandy Street, Exeter EX4 3LS, UK, 1997.

22 N. G. Bisset, M. Wichtl (Editors), 'Herbal Drugs and Phytopharmaceuticals: a Handbook for Practice on a Scientific Basis', CRC Press, Ann Arbor, USA, 1994.

23 A. Chevallier, 'The Encyclopedia of Medicinal Plants', Dorling Kindersley, London, 1996.

24 M. G. L. Hertog, E. J. M., Feskens, P. C. H. Hollman, *et al.*, *Lancet*, 1993, **342**, 1007.

25 A. Weikl, H. S. Noh, *Herz. Gefabe.*, 1993, **11**, 516.

26 C. A. Newall, L. A. Anderson, J. D. Phillipson, 'Herbal Medicines: a Guide for Health-care Professionals', The Pharmaceutical Press, London, 1996.

27 G. Marakis, Transfer Report for PhD, The University of Reading, 1998.

28 H. P. Koch, L. D. Lawson (Editors), 'Garlic: the Science and Therapeutic Application of *Allium sativum* L. and Related Species', 2nd edition, Williams & Williams, Baltimore, 1996.

29 British Herbal Pharmacopoeia, British Herbal Medicine Association, PO Box 304, Bournemouth, BH7 6JZ, UK, 1983.

Plant Sterol-enriched Margarines: Efficacy in Cholesterol Lowering and Effects on Lipid Soluble Vitamins

T. van Vliet,[1] G.W. Meijer,[2] J.A. Weststrate[2] and
H.F.J. Hendriks[1]

[1] TNO NUTRITION AND FOOD RESEARCH INSTITUTE, PO BOX 360,
3700 AJ ZEIST, NETHERLANDS
[2] UNILEVER NUTRITION CENTER, UNILEVER RESEARCH,
VLAARDINGEN, NETHERLANDS

1 Introduction

1.1 Plant Sterols

Plant sterols, also called phytosterols, are naturally occurring plant compounds, structurally related to cholesterol, but different in their side-chain configuration. The most predominant plant sterols in nature are β-sitosterol, campesterol and stigmasterol. In nature, plant sterols occur primarily in the unhydrogenated free form. The hydrogenation of phytosterols forms saturated phytosterols, such as sitostanol and campestanol. Sterols are only approximately 2% soluble in oil while they are virtually insoluble in water. Solubility can be improved by esterification.

The most important dietary sources in humans are corn, bean and plant oil and oil-based products such as margarines. Plant sterol intake of non-vegetarian adults ranges between 3 and 6 mg/kg body weight/day. In vegetarians intakes are about 40% higher.[1]

Plant sterols are not synthesized in the human body. Their intestinal absorption is less than 5% of dietary levels and plasma levels are generally less than 1% of cholesterol levels. Absorbed phytosterols are efficiently cleared from the circulation *via* the biliary route.[1] Their poor absorption is not fully understood. One proposed explanation is that phytosterols may be inadequately esterified in the intestine.[1]

A great number of studies with plant sterol intakes up to 25 g/d for several months have been performed without reporting side effects.[2] Only patients with sitosterolemia, a rare inherited lipid storage disease leading to premature atherosclerosis, have a very high absorption rate of phytosterols and show serious side effects.[1]

1.2 Reported cholesterol lowering properties in human nutrition studies

The cholesterol lowering properties of plant sterols have been known since the 1950s.[2] The mechanisms of hypocholesterolaemic action are not fully understood, but probably include inhibition of intestinal cholesterol absorption. In addition, plant sterols reaching the liver may alter cholesterol synthesis and excretion.[1]

The original literature on cholesterol reduction by phytosterols, concerning high dosages applied as drug in hypercholesterolaemic patients, has been reviewed by Pollak and Kritchevski.[2] More recently, a number of human studies have been done with normo- or mildly hypercholesterolaemic subjects, using relatively low dosages of plant sterols, either in capsules or in a spread or mayonnaise. An overview of those studies with normo- or mildly hypercholesterolaemic, otherwise healthy, subjects is given in Table 1. In addition, a number of studies have been reported with diabetic patients (not included in the table).

Only in the study of Denke[6] was no effect reported. The author suggested, that is was possibly because the subjects were on a strict cholesterol lowering diet. The other reported studies found reductions in total cholesterol of 7 to 15%; these variations seem not to be explained by the dosage used or the duration of treatment.

2 Effects of Various Plant Sterol-enriched Margarines

2.1 Introduction

Sterols with different chemical structures may vary in their potential to reduce plasma cholesterol levels. Most reported studies used sitostanol-esters, and no comparison in cholesterol lowering efficacy between sitostanol-ester and other sterols has been reported. Therefore a study was done to assess, in normocholesterolaemic and mildly hypercholesterolaemic subjects, the efficacy to lower plasma total- and LDL-cholesterol concentrations of margarines enriched with sterols derived from various oils, and to compare this efficacy with that of sitostanol-ester. The study is described in more detail in reference 8.

2.2 Subjects and Methods

The study was conducted at the Unilever Nutrition Centre in Vlaardingen, The Netherlands. The study protocol was approved by the Medical and Ethical Committee of Unilever Nederland BV.

Table 1   *Overview of the main studies on cholesterol lowering properties of plant sterols in normo- or hypercholesterolaemic, otherwise healthy, humans*

Subjects & Design	Treatment	Cholesterol	Ref
n = 6 Hypercholesterolaemic Chol. 7.6–13.6	4 wk, 1.5 g/d sitostanol (88%) as capsules	Tot: 15% LDL: 15% HDL: uneff.	3
n = 67 (47, 20) Hypercholesterolaemic Chol. > 6.0* Parallel	6 wk, 3.4 g/d Sitostanol ester in rapeseed oil mayonnaise (to replace 50 g of visible fat)	Tot: 7.5% LDL: 10% HDL: uneff.	4
N = 153 (42%) Hypercholesterolaemic Chol. > 5.58, mean 6.1 Parallel (3 groups)	1 year control or 2.6 g/d or $\frac{1}{2}$ year 2.6 followed by $\frac{1}{2}$ year 1.8 g/d Sitostanol ester in margarine	Tot: 10.2% LDL: 14.1%	5
n = 33 (males) Hypercholesterolaemic Chol. mean = 6.18⁋	3 months diet⁋ + 3 g/d Sitostanol as capsules Intake during 3 meals	No effects	6
n = 12 Normolipidaemic Chol mean 4.30 Cross-over	4 wk, 0.74 g/d Soybean phytosterols in butter	Tot: 10% LDL: 15%	7

* After the 4-week run-in period with rapeseed oil, average total cholesterol was 5.9 mmol/L.
⁋ After a diet containing < 200 mg cholesterol per day

2.2.1 Margarines and Treatments. The oils used were soybean oil, ricebran oil and sheanut oil, and the margarines enriched with those oils were compared to the non-enriched margarine 'Flora' (Van den Bergh Foods, Crawley, UK) and a margarine fortified with sitostanol-ester 'Benecol' (Raisio Inc, Finland). The margarines were used in amounts of 30 g per day, providing 0.1 g sterols per day for Flora, 2.7 for Benecol, 3.2 for soybean oil sterol enriched margarine, 1.7 for ricebran oil sterol enriched margarine and 3.0 g sterol per day for sheanut oil sterol enriched margarine.

2.2.2 Subjects and Design. One hundred volunteers, 50 males and 50 females, participated in the study. All volunteers gave their written informed consent for participation prior to the start of the study. Participants were all apparently healthy. Baseline total and LDL-cholesterol levels were 5.35 ± 1.06 and 3.54 ± 0.97 mmol/L, respectively.

The study had a randomized double-blind placebo-controlled balanced incomplete Latin square design with five treatments and four periods. Subjects received, in four consecutive periods of 24 or 25 days (3.5 weeks), 30 g per day of a margarine in a coded tub for consumption at lunch and dinner, but not for cooking. The margarines were meant to replace an equivalent amount of the spreads habitually used by the volunteers.

Effects of margarine consumption were assessed on plasma lipids, routine blood chemistry and haematology, plasma α- plus β-carotene and lycopene concentrations, plasma phytosterol concentrations, fatty acid composition of plasma cholesteryl esters and on habitual energy, total fat, fatty acid and cholesterol intake.

Statistical evaluation of treatment effects was done using analysis of variance, using the following factors: gender, subject, period, treatment, carry-over, diet × gender.

2.3 Results

2.3.1 General. A total of 95 subjects completed the study. Compliance with the treatment was excellent. Average margarine consumption varied between 29.8 g/d for the ricebran sterol margarine to 30.4 g/d for the sheanut sterol margarines. Routine blood chemistry and haematology showed no evidence for adverse effects of consumption of the sterol enriched margarines.

2.3.2 Plasma Lipids. Figure 1 shows blood cholesterol data. Benecol and the soybean margarine significantly lowered total and LDL-cholesterol concentrations by about 0.37–0.44 mmol/L compared to Flora, without an effect on HDL-cholesterol concentration. The sheanut or ricebran sterol margarines did not differ in their effect on blood lipids compared to Flora. Total glycerol concentrations were not affected by any treatment. There was no significant difference between males and females in treatment effect.

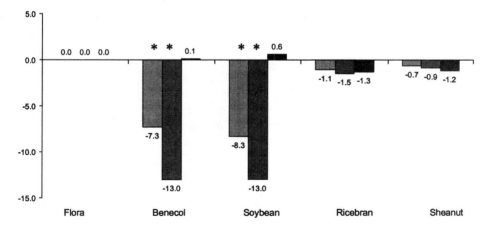

Figure 1 *Changes of total, LDL- and HDL-cholesterol concentrations after consumption of the sterol-enriched margarines as compared to Flora (%). * Indicate significant differences as compared to Flora (p<0.05).*

Treatment effects were very similar for normocholesterolaemic subjects (*i.e.* the lowest tertile of cholesterol concentrations at study-entry) and mildly hypercholesterolaemic subjects (*i.e.* the highest tertile of cholesterol concentrations at study-entry).

2.3.3 Plasma Carotenoids. Figure 2 shows plasma carotenoid data after standardisation for concentrations of plasma lipids. All sterol fortified margarines caused a lowering of blood carotenoid concentrations, but the decrease in lycopene concentration was not statistically significant for the soybean margarine and the ricebran margarine.

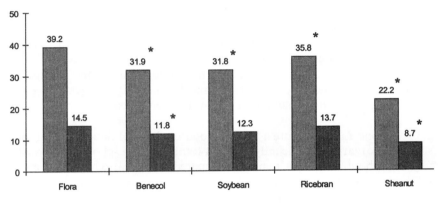

Figure 2 *Lipid adjusted plasma (α+β)-carotene and lycopene concentrations after consumption of the margarines (in µg/mmol lipids). * Indicate significant differences as compared to Flora (p<0.05).*

2.4 Conclusions

Margarines containing sheanut or ricebran oil sterols were not effective in lowering blood cholesterol concentrations compared to the control margarine. Margarine with sterol-esters from soybean oil reduces plasma total and LDL-cholesterol concentrations by 8.3 and 13.0%, respectively, as compared to control margarine, and is as effective as margarine with sitostanol-ester. HDL-cholesterol concentrations were not affected by any of the margarines. The soybean oil sterol enriched margarine has the same effect in normo- as in mildly hypercholesterolaemic subjects. All sterol enriched margarines tested reduced blood carotenoid concentrations somewhat.

3 Effects of Three Intake Levels of Vegetable Oil Sterols

3.1 Introduction

Plant sterols were found to reduce plasma cholesterol concentrations. However, they may also lower plasma concentrations of other lipophilic compounds. Therefore, an optimal plant sterol intake level has to be selected, reducing cholesterol concentrations optimally but having minimal effect on other lipophilic compounds. The main objective of the study described in this section was to investigate the dose-dependency of the cholesterol-lowering effects of three dosages of plant sterols in spreads. The study is described in more detail in reference 9.

3.2 Subjects and Methods

The study was conducted according to Good Clinical Practice at the TNO Nutrition and Food Research Institute, Zeist, The Netherlands. The study protocol was approved by the TNO Medical Ethical Committee.

3.2.1 Spreads. The five spreads used included three test spreads, which were spreads fortified with three concentrations of plant sterols derived from vegetable oil (predominantly soybean oil), butter, and a control spread, 'Flora' (Van den Bergh Foods, Crawley, UK), with a fatty acid composition very similar to that of the test spreads. The three test spreads provided about 0.8, 1.6 and 3.2 g of plant sterols per day.

3.2.2 Subjects and Design. One hundred volunteers, 42 males and 58 females, participated in the study. All volunteers gave their written informed consent for participation prior to the start of the study. Participants were all apparently healthy. Baseline total and LDL-cholesterol concentrations were 5.10 ± 0.97 (range: 2.7–7.4) and 2.97 ± 0.83 (range: 1.12–5.22) mmol/L, respectively.

The study had a randomized double-blind placebo-controlled balanced incomplete Latin square design using five spreads and four periods. This design controls for between subject variation and for variation over time. Subjects received, in four consecutive periods of 24 or 25 days (3.5 weeks), 25 grams of spread per day for consumption at lunch and dinner, but not for cooking. The spreads were meant to replace an equivalent amount of the spreads habitually used by the volunteers.

Effects of spread consumption were assessed on plasma lipids, routine blood chemistry and plasma lipid soluble (pro)vitamins. Additional parameters measured were body weight, dietary intake, assessed with a food frequency questionnaire at the end of each period, and adverse events.

Statistical evaluation of treatment effects was done using analysis of variance, using the following factors: subject, spread, period and residual effects (including carry-over).

3.3 Results

3.3.1 General. All subjects completed the study. Compliance with the treatment was generally very good, only about 1% of the portions of spread was not consumed, distributed equally over the five spreads. Nutrient intake, excluding spread intake, was calculated per treatment period. Total fat intake and the contribution of saturated, monounsaturated and polyunsaturated fatty acids to fat intake, dietary cholesterol and energy intake did not change during consumption of spreads, nor did any other calculated dietary intake parameter. Adverse event reporting and analyses of liver enzymes did not show side effects of any of the spreads applied.

3.3.2 Plasma Lipids. Figure 3 shows the percentage change in blood cholesterol as compared to Flora. Total and LDL-cholesterol concentrations were significantly decreased by plant sterol consumption, whereas HDL-cholesterol and triacylglycerol concentrations were not affected. Total and LDL-cholesterol concentrations decreased after Flora consumption as compared to butter consumption. Decreases in total cholesterol after intake of the sterol-enriched spreads as compared to Flora were 0.26 (95% confidence interval, CI: 0.15–0.36), 0.31 (CI:0.20–0.41) and 0.35 (CI:0.25–0.46) mmol/L, for daily consumption of 0.8, 1.6 and 3.3 g plant sterols, respectively. For LDL-cholesterol these decreases were 0.20 (CI:0.10–0.31), 0.26 (CI:0.15–0.36) and 0.30 (CI:0.20–0.41) mmol/L, respectively. Differences in cholesterol reductions between the plant sterol doses consumed were not significant, although 95% confidence intervals (as compared to control) suggest increasing cholesterol reductions with increasing plant sterol content. However, also no significant trend in any of the serum lipids analysed could be demonstrated.

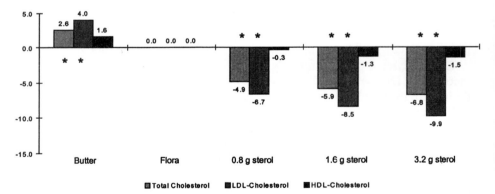

Figure 3 *Changes of total, LDL- and HDL-cholesterol concentrations after consumption of the spreads as compared to Flora (%). * Indicates significant differences as compared to Flora (p<0.05).*

3.3.3 Plasma Lipid-soluble (Pro)vitamins. Plasma concentrations of vitamin K1 and 25-OH-vitamin D were not affected by consumption of the spreads enriched in plant sterols. The compounds (α+β)-carotene, lycopene and α-tocopherol were decreased after consumption of plant sterol enriched spreads as compared to Flora consumption. After lipid standardization, to correct for the decreases in plasma lipids, only the decreases in (α+β)-carotene after intake of 0.8 and 3.2 g plant sterols per day were statistically significant. Figure 4 shows plasma carotenoid data.

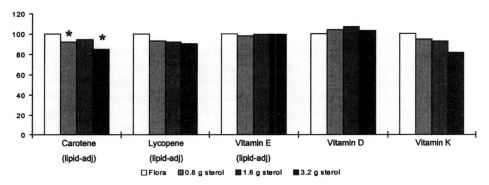

Figure 4 *Plasma lipid soluble (pro)vitamin concentrations after consumption of test spreads as compared to Flora (100%). * Indicates significant differences as compared to Flora (p<0.05).*

3.4 Conclusions

Daily intake of spreads enriched with 0.8–3.2 g plant sterols, predominantly derived from soybean oil, decreased total cholesterol by 4.9–6.8% as compared to consumption of a control spread. LDL-cholesterol was decreased by 6.7–9.9% as compared to the control spread. The test spreads did not affect HDL-cholesterol as compared to the control spread. Plasma vitamin K1 and 25-OH-vitamin D and lipid standardized plasma α-tocopherol and lycopene were not affected by the plant sterol enriched spreads. However, lipid standardized plasma (α+β)-carotene was reduced by daily consumption of 0.8 and 3.2 g plant sterols.

4 Overall Conclusions and Implications for Public Health

4.1 Overall Conclusion

Plant sterols, predominantly derived from soybean oil, in amounts of 0.8 to 3.2 g per day are effective in lowering total cholesterol by 5–8% and LDL-cholesterol by 7–13%, without affecting HDL-cholesterol. Lipid standardized plasma ($\alpha+\beta$)-carotene is reduced by 6–19% by daily consumption of 0.8 and 3.2 g plant sterols, but unaffected by daily consumption of 1.6 g plant sterols.

4.2 Implications for Public Health

Cardiovascular diseases (CVD) are the major cause of deaths in most developed countries. About half of all CVD deaths are from coronary heart disease (CHD), while about one quarter are from stroke. One of the major modifiable risk factors for coronary heart disease (CHD) mortality is plasma cholesterol concentration. Several large scale intervention trials have shown that cholesterol lowering drug therapy in hypercholesterolaemic patients beneficially affects CHD mortality risk.[10] However, most people who die from CHD have only moderately raised cholesterol concentrations. Mildly hypercholesterolaemic subjects usually have dietary guidelines prescribed as a first treatment to lower their blood cholesterol. However, the percentage reduction in blood total cholesterol attributable to dietary advice is modestly effective in free-living subjects; that is a reduction of total cholesterol between 3 and 6%.[11] Cholesterol reductions observed in the studies described above with vegetable oil sterol enriched margarines, of 5–8% for total cholesterol, would decrease CHD risk by about 25–40% at age 40 y and by about 10–15% at age 70 y.[12] When comparing with butter consumption, even higher risk reductions are obtained. These are large effects which, on a population basis, would substantially contribute to the prevention of cardiovascular disease.

However, in addition to the cholesterol lowering effect, the vegetable oil sterol enriched margarines reduce lipid standardized plasma ($\alpha+\beta$)-carotene somewhat. Carotenoids may have positive effects on health[13–15] and, therefore, minimization of the carotenoid lowering effect of plant sterols should be considered.

References

1   W.H. Ling and P.J.H. Jones, *Life Sci.*, 1995, **57**, 195.
2   O.J. Pollak and D. Kritchevsky, 'Sitosterol'. Monographs on atherosclerosis, vol. 10. Karger, Basel, 1981.
3   T. Heinemann, O. Leiss and K. Van Bergmann, *Atherosclerosis*, 1986, **61**, 219.
4   H.T. Vanhanen, S. Blomqvist, C. Ehnholm, M. Hyvönen, M. Jauhiainen, I. Torstila and T.A. Miettienen, *J. Lipid Res.*, 1993, **34**, 1535.
5   T.A. Miettinen, P. Puska, H. Gylling, H. VanHanen and E. Vartiainen, *N. Engl. J. Med.*, 1995, **333**, 1308.
6   M.A. Denke, *Am. J. Clin. Nutr.*, 1995, **61**, 392.
7   X. Pelletier, S. Belbraouet, D. Mirabel, F. Mordret, JL Perrin, X. Pages and G. Debry, *Ann. Nutr. Metab.*, 1995, **39**, 291.
8   J.A. Weststrate and G.W. Meijer, *Eur. J. Clin. Nutr.*, 1998, **52**, 334.
9   H.F.J. Hendriks, J.A. Weststrate, T. Van Vliet and G.W. Meijer, *Eur. J. Clin. Nutr.*, 1999, **53**, 319.
10   A.L. Gould, J.E. Rossouw, N.C. Santanello, J.F. Heyse and C.D. Furberg, *Circulation*, 1998, **97**, 946.

11.   J.L. Tang, J.M. Armitage, T. Lancaster, C.A. Silagy, G.H. Fowler and H.A.W. Neil, *B.M.J.*, 1998, **316**, 1213.
12.   M.R. Law, M.J. Wald and S.G. Thompson, *B.M.J.*, 1994, **308**, 367.
13.   G. van Poppel, *Eur. J. Cancer*, 1993, **29A**, 1335.
14   S.T. Mayne, *FASEB J.*, 1996, **10**, 690.
15   J.E. Manson, J.M. Gaziano, M.A. Jonas and C.H. Hennekens, *J. Am. Coll. Nutr.*, 1993, **4**, 426.

Red Clover Isoflavone Supplementation in Peri-menopausal Management

A.J. Husband

NOVOGEN LIMITED, 140 WICKS ROAD, NORTH RYDE, NSW, 2113, AUSTRALIA

1 Introduction

Throughout human evolution, oestrogenic influences in the body have involved a balance between steroidal oestrogenic hormones, and oestrogenic substances derived from plants, especially the class of plant phenolic compounds known as isoflavones. Legumes are the richest source of these compounds and the traditional dependence on legumes as a source of dietary protein has permitted an evolutionary adaptation by which steroidal oestrogens in women cease to be produced after reproductive years, protecting the breast and uterus from constant stimulation, after which plant oestrogens remain as the source of oestrogenic support for non-reproductive tissues. However, in recent times, western cultures have turned away from legumes as a protein source with a preference instead for protein from animal sources, such as meat and dairy products. This has led to an imbalance in oestrogenic activity and the emergence of symptoms of oestrogen deficiency, especially after the menopause in western women.

The proposition that oestrogen deficiency disorders may result from a dietary imbalance, associated with western dietary habits, suggests two options for management – implementation of a major change in dietary habits, or provision of oestrogen supplementation. The latter is the favoured option for western women but supplementation with steroidal oestrogens carries increased risk of breast and uterine cancer[1] and cardiovascular complications due to thrombogenic effects.[2]

It has been proposed that dietary supplementation with isoflavones, the natural plant oestrogens present in traditional diets, may provide the benefits of steroidal oestrogen hormone replacement therapy (HRT) without the risks.[3] This paper presents data obtained from clinical trials, which indicate that 1 tablet per day of a standardised supplement of red clover isoflavones (40 mg) in tablet form is sufficient to maintain circulating isoflavone levels equivalent to those achieved with traditional legume-rich diets, and that this level of supplementation can alleviate the acute vasomotor symptoms of menopause (hot flushes), as well as correct deterioration in cardiovascular function, one of the long term effects of oestrogen deficiency. Evidence will also be presented demonstrating that a higher level of supplementation with red clover isoflavones can also maintain beneficial blood cholesterol profiles.

2 Red Clover Isoflavone Supplement

The studies described in this paper have been performed using standardised supplements of red clover isoflavones in tablet form. While soy isoflavone extracts have been used in many studies of isoflavone supplementation, these supplements only contain two isoflavones, daidzein and genistein, but red clover has the advantage of providing all four of the important isoflavones, namely daidzein and genistein as well as their respective methylated forms, formononetin and biochanin. The studies described in this paper were performed either with Novogen Standard Red Clover Extract, formulated as standardised 500 mg tablets containing 40 mg total isoflavones, or with Novogen P-081 also formulated as standardised 500 mg tablet but containing 25 mg total isoflavones with a modified isoflavone ratio profile.

3 Pharmacokinetics of Isoflavone Supplementation with Red Clover Extract

Despite the widespread use of dietary supplements containing isoflavone phytoestrogens to prevent problems associated with oestrogen deficiency, there is little information about their pharmacokinetics in humans.

Thus a study was performed by Howes and colleagues (St George Hospital, Sydney) to measure the plasma and urinary profiles of isoflavones after acute and chronic administration of Novogen Standard Red Clover Extract tablets.[4]

Fourteen subjects, who had been consuming a low isoflavone diet for two weeks were given an oral dose of one tablet of Novogen Standard Red Clover Extract at 0900 h after an overnight fast. Venous plasma and urine were collected at intervals for 24 h, after which subjects commenced taking two tablets each day for two weeks and repeated plasma samples and urine were collected for a 48 h period after the last dose. Plasma and urine isoflavones were assayed by HPLC.

After the acute dosing, all four isoflavones appeared rapidly in plasma and reached peak levels at around 5–6 h. Although daidzein and genistein were present in much higher concentrations than their methylated precursors, indicating a rapid demethylation of these compounds, formononetin and biochanin remained detectable in both plasma and urine at all times. Since formononetin and biochanin have a range of biological effects additional to daidzein and genistein, this demonstrates the potential for supplements containing all four isoflavones to produce a broader range of physiological effects than soy supplements which contain only the demethylated forms.

Dose-corrected peak and trough plasma levels were higher following chronic than following acute dosing, indicating accumulation of isoflavones in plasma during chronic therapy. From plasma concentrations of isoflavones during chronic isoflavone tablet administration it was calculated that, after dose correction, the circulating levels achieved with 1 tablet per day would be similar to those found in populations consuming diets with a high isoflavone content.

These data demonstrate that once daily administration of Novogen Standard Red Clover Extract tablets can achieve plasma isoflavone levels equivalent to those reported for populations consuming diets with a high isoflavone content.[5,6]

4 Effects on Acute Menopause Symptoms

In an open trial conducted by Nachtigall and colleagues (New York University),[7] 16 out of 20 women reported improvement in menopause symptoms after taking 1 tablet/day of Novogen Standard Red Clover Extract tablets for 12 weeks.

To confirm this finding that increasing the intake of isoflavones by dietary supplementation may produce a therapeutic effect in reducing the incidence and severity of hot flushes in menopausal women, a placebo controlled trial was undertaken by Dr R Baber (Royal North Shore Hospital, Sydney). In this study, 51 postmenopausal women were randomised to placebo and active (1 tablet per day of Novogen Standard Red Clover Extract, a 40 mg isoflavone supplement) in a cross-over design trial.[8] After a 1 week run-in period, subjects were commenced on a 12 week period of treatment (active or placebo), followed by a 1-month placebo washout period, then crossed over to the alternate treatment regime for a further 14 weeks. Symptom diaries were maintained throughout the trial and at the start and end of treatment plasma SHBG assay, full blood count, biochemical profiles, vaginal swabs and vaginal ultrasounds were performed and isoflavones determined in 24 h urine collections by HPLC analysis.

There was no significant difference between active and placebo groups in the reduction in hot flushes between start and finish time points. However, analysis performed on data obtained at interim time points revealed that, in the active group, there was a reduction in flushing incidence, relative to incidence at baseline, of 17%, 23% and 21% at 4, 8 and 12 weeks, respectively, after commencement of treatment. In placebo controls there was a reduction of only 3%, 2% and 17% at the same time points. However, owing to the high variability between patients, differences between active and placebo at each time point were not statistically significant. Many of the placebo subjects had high levels of urinary excretion of isoflavones despite advice to avoid known food sources. This appeared to explain the high placebo response at week 12. There were no significant differences between groups for haematological or biochemical parameters and vaginal swab or uterine ultrasound findings, indicating that three months of isoflavone supplementation did not cause adverse events or endometrial changes.

The combined values for all subjects, regardless of treatment group, revealed a strong negative correlation between the level of urinary isoflavone excretion and the incidence of hot flushes. These data demonstrated that any subjects with urinary isoflavone excretion greater than 2 mg/day reported reduction in flushing greater than the placebo response. Since the preceding pharmacokinetic study demonstrated that one Novogen Standard Red Clover Extract tablet per day was sufficient to achieve isoflavone levels of this magnitude, this prompts the conclusion that isoflavones reduce flushing symptoms whether derived from diet or from supplementation. These data further demonstrate that for those whose diets were deficient in isoflavones, supplementation using Novogen Standard Red Clover Extract tablets provided sufficient isoflavone input to obtain a reduction in flushing. These data also indicate that part of the placebo effect observed in many studies of menopause symptoms may be attributable to dietary sources of isoflavones.

5 Effects on Vascular Compliance

Poor vascular compliance (blood vessel elasticity) is associated with an increased risk of stroke and heart attack. Impaired vascular compliance has been shown to associate with ageing, menopause, hypertension and possibly hyperlipidemia, and probably to be partly responsible for systolic hypertension in elderly people. It has been shown previously that steroidal oestrogens reduce the prevalence of coronary heart disease in menopausal women[9] and untreated menopausal women have endothelial dysfunction and reduced carotid arterial pulsatility.[10,11]

The possibility that the heightened cardiovascular risk associated with the menopause can be reduced with isoflavones, was tested in a study conducted by Nestel and colleagues (Baker Medical Research Institute, Melbourne) by measuring arterial compliance, an index of the elasticity of large arteries such as the thoracic aorta, in a placebo controlled trial in postmenopausal women.[12]

An initial 3–4 week run-in period and a 5 week placebo period were followed by two 5 week periods of active treatment with 1 tablet per day (40 mg dose) and then 2 tablets per day (80 mg dose) of Novogen Standard Red Clover Extract in 14 and 13 women respectively, with 3 others serving as placebo-controls throughout. Arterial compliance, measured by ultrasound as a pressure (carotid artery) and volume (outflow into aorta) relationship, was measured after each period; plasma lipids were measured twice during each period.

Arterial compliance rose by 23% with the 80 mg isoflavone dose and slightly less with the 40 mg dose (placebo: 18.6 ± 1.50; 40 mg: 23.7 ± 1.50; 80 mg: 24.4 ± 1.37). In the 3 women taking continuous placebo, compliance was 16 ± 2.2, *i.e.* similar to the run-in period for all: 17 ± 2.1. ANOVA showed a significant (p=0.009) difference between treatments; by Bonferroni multiple comparison differences between placebo (or run-in) versus both 80 mg and 40 mg isoflavone doses were significant. This effect was equivalent to that reported in other studies using oestrogen hormone replacement therapy (HRT).[13]

Thus an important cardiovascular risk factor, arterial compliance, which diminishes with the menopause, was significantly improved with red clover isoflavones. Since diminished compliance leads to systolic hypertension and may increase left ventricular work, the findings indicate a potential new therapeutic approach to reduce cardiovascular risk in postmenopausal women without resorting to steroidal oestrogen therapy.

6 Effects on Blood Cholesterol

In a study conducted by Eden *et al.*[14] an increase in plasma HDL cholesterol of 18.1% was observed in patients taking 1 tablet/day Novogen Standard Red Clover Extract. However, this effect was not observed in the study reported above.[12] Since it was possible that for reliable lipid modification a higher dose or different ratio of isoflavones than that obtained from Novogen Standard Red Clover Extract was required to achieve lipid profile modification, a new preparation, Novogen P-081, in which the ratio of the four isoflavones is modified, was tested in postmenopausal women. This study, conducted by Dr R Baber and Dr P Clifton-Bligh at the Royal North Shore Hospital in Sydney, was designed to determine the effects of this supplement on both lipid and bone metabolism in postmenopausal women.

The study design involved 60 postmenopausal women randomised to 25, 50 or 75 mg Novogen P-081. After a 4 week run-in period, all subjects commenced a 24 week treatment period, followed by 8 weeks placebo, then an open phase in which all subjects were maintained on 25 mg P-081 for 26 week. Blood, 24 h urine, bone density, mammogram and full blood count were performed at the start and end of the treatment phase, and a uterine ultrasound was performed at the end of the treatment and open phases. Although the trial is not yet complete and the data are still coded, preliminary analysis of the lipid data revealed that, regardless of the dose level during the treatment phase, all subjects displayed elevated HDL cholesterol levels relative to their baseline measurements at 3 months, but more pronounced effects were observed at 6 months (Figure 1). The effects on lipid profile of each of the 3 dose rates as well as effects on bone mineral density have not yet been analysed but a full analysis will be performed when all subjects have completed the protocol.

Evidence obtained by Lyons-Wall and colleagues (University of Sydney) indicates that favourable modification of lipid profiles may also be achieved in premenopausal women if a sufficiently high dose of red clover isoflavones is provided. In a preliminary study of premenopausal women, no effects on lipid profile were observed using 1 tablet per day of 40 mg isoflavone supplement. In a subsequent study,[15] women aged between 24 and 37 years of age were randomised to active or placebo groups and observed over 4 menstrual cycles

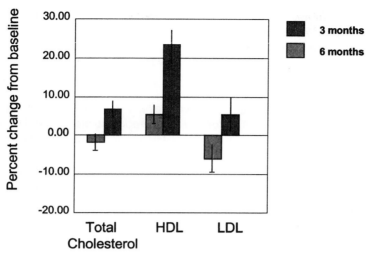

Figure 1 *Percentage change in lipid fractions in subjects taking 25, 50 or 75 mg/day Novogen P-081*

(6 women per group). The active treatment group were given 2 placebo tablets per day during the first cycle then 2 tablets per day Novogen Standard Red Clover Extract for 3 cycles. Women in the placebo group received 2 placebo tablets per day for 4 cycles. An increase in HDL cholesterol with no change in LDL cholesterol was observed at the end of the study period in each of the 6 women in the active group (mean 12% increase in HDL). No changes were observed in any of the control group women.

7 Discussion and Conclusions

The data presented here demonstrate the potential for red clover isoflavone supplementation to provide protection against the effects of oestrogen deficiency, arising as a result of the menopause in alleviating the acute vasomotor symptoms of menopause (hot flushes) as well as maintaining cardiovascular health as a result of improved vascular compliance and improved cholesterol profile. These effects were obtained in the absence of any of the recognised side effects attributable to steroidal oestrogen HRT treatments. The benefits of isoflavone supplementation are most pronounced in subjects with low isoflavone intake but, if appropriate dietary modification is exercised, similar benefits can be obtained.

Although they are weak oestrogens, phytoestrogens are increasingly being recognised as an important and natural source of complementary oestrogens for maintaining human hormone balance. In previous attempts to use phytoestrogens to treat menopause symptoms there have been problems in ensuring sufficient daily intake of plant oestrogens from dietary modification to achieve a biological effect because of low concentration in food sources, and plant extracts have been poorly standardised leading to variable responses. The only other study reporting statistically significant beneficial effects of isoflavone supplementation on menopause symptoms was reported by Albertazzi *et al*,[16] in which a dose of approximately 60 mg soy isoflavones was required to achieve a reduction in flushing of 45% versus a 30% reduction obtained with placebo. Lipid modification has also been reported using soy isoflavones but a dose of 56 mg per day was required to achieve an 11% increase in HDL cholesterol.[17] The greater efficacy of red clover isoflavone supplementation is due in part to the fact that all four of the important dietary isoflavones are present in red clover (whereas most other legumes contain only two) and that the supplements used in the studies reported in this paper were made available in a standardised tableted dose form. This now represents a credible alternative to steroidal oestrogen supplementation although whether the two approaches may be used concurrently also requires investigation.

References

1   A. Cooper and M. Whitehead, *Current Opinion in Obstetrics & Gynecology*, 1995, **7**, 214.

2   S. Hulley, D. Grady, T. Bush, C, Furberg, D. Herrington, B. Riggs and E. VIittinghoff, *Journal of the American Medical Association*, 1998, **280**, 605.

3   T.B. Clarkson, M.S. Anthony, J.K. Williams, E.K. Honore and J.M Cline, *Proceedings of the Society for Experimental Biology & Medicine*, 1998, **217**, 365.

4   J.B. Howes, J.A. Eden, M.A. Waring, L. Huang and L.G. Howes,. *Climacteric*, 1999 **(In Press)**.

5   W.M. Mazur, J.A. Duke, K. Wahala, S. Rasku and H. Adlercreutz, *IInd International Symposium on the Role of Soy in Preventing and Treating Chronic Diseases*, Sept 15–18, 1996.

6   H. Adlercreutz, H. Markkanen and S. Watanabe, *Lancet*, 1993, **342**, 1209.

7   L.B. Nachtigall, R. Fenichel, L. Lagrega, W.W. Lee, M.J. Nachtigall and L.E. Nachtigall, *The Female Patient*, (1998) **(In Press)**.

8   R.J. Baber, C. Templeman, T. Morton, G.E. Kelly and L. West, *Climacteric*, (1999) **(In Press)**.

9   T.L. Bush, E. Barrett-Connor, L.D. Cowan, M.H. Criqui, R.B. Wallace, C.M. Suchindran, H.A. Tyroler and B.M. Rifkind, *Clinical Investigation*, 1987, **75**, 1102.

10   S. Taddei, A. Virdis, L. Ghiadoni, P. Mattei, I. Sudano, G. Bernini, S. Pinto and A. Salvetti, *Hypertension*, 1996, **28**, 576.

11   K.F. Gangar, S. Vyas, M. Whitehead, D. Crook, H. Meire and S. Campbell, *Lancet*, 1991, **338**, 839.

12   P.J. Nestel, S. Pomeroy, S. Kay, P. Komesaroff, J. Behrsing, J.D. Cameron and L. West, *Journal of Clinical Endocrinology and Metabolism*, 1999, **84**, 895.

13   B.P. McGrath, Y-Y. Liang, H. Teede, L.M. Shiel, J.D. Cameron and A. Dart, *Arterioscler. Thromb. Vasc. Biol.*, 1998, **18**, 1149.

14   D.C. Knight, J.B. Howes and J.A. Eden, The effect of Promensil, *Climacteric*, (1999) **(In Press)**.

15   S.J. Smith, P.M. Lyons-Wall and S. Samman, Effect of an isoflavone supplement on cholesterol metabolism in premenopausal women – a pilot study. Abstract, Proceedings Nutrition Society of Australia Annual Meeting, Adelaide, 1998.

16   P. Albertazzi, F. Pansini, G. Bonaccorgsi, L. Zanotti, E. Forini and D. De Aloysio, *Obstet. Gynecol.*, 1998, **91**, 6.

17   J.A. Baum, H. Teng, J.W.J. Erdman, R.M. Weigel, B.P. Klein, V.W. Persky, S. Freels, P. Surya, *et al.*, *American Journal of Clinical Nutrition*, 1998, **68**, 545.

Phyto-oestrogens in Neonatal Diets

A.B. Hanley, S.L. Oehlschlager, J. McBride, B. Pöpping,
R. Smith, A. Damant, K. Barnes and M. Fewtrell*

CSL FOOD SCIENCE LABORATORY, NORWICH RESEARCH PARK,
COLNEY, NORWICH, NR4 7UQ, UK
*MRC CHILDHOOD NUTRITION RESEARCH CENTRE, INSTITUTE
OF CHILD HEALTH, LONDON, WC1N 1EH, UK

1 Introduction

Plant-derived oestrogens have been consumed for thousands of years. The greatest single source of exposure for most individuals is soya, which contains a range of structurally similar isoflavones (Figure 1). The major structural characteristics which appear to be associated with oestrogenicity are the phenol ring and hydroxy substituents.[1,2] Much of the oestrogenic activity of genistein and daidzein (the major soya phyto-oestrogens) would appear to be associated with binding to the oestrogen receptor however non-receptor mediated effects have also been noted.

The oestrogen receptor (ER) occurs in two forms (α and β) and appears to exhibit some degree of tissue specificity.[3,4] Isoflavones have a stronger binding affinity for ER β than for ER α[4] and this may suggest a differential effect which will compound the normal variation in bioavailability for different tissues.

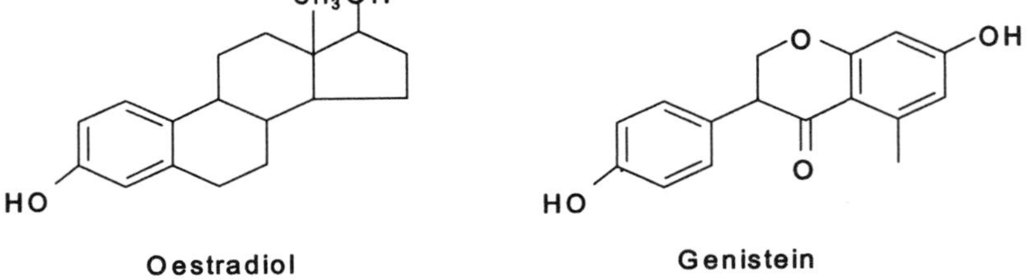

Figure 1 *Structures of oestradiol and genistein*

The biological effects of the oestrogen receptor is associated with the conversion of the receptor to an active form after dimerisation and binding of the ligand.[5] The active receptor contains a DNA-binding domain and a transactivating region whose purpose is to cause specific genes to be transcribed. The target genes are those which are associated with a specific sequence called an oestrogen responsive element. It has recently been demonstrated that both agonists and antagonists bind at the same site in the receptor. However, they appear to induce different conformations in the transactivating domain[6] and this may help to explain differential effects.

While mechanistic studies which relate specific compounds to measurable biological events are beginning to be carried out, a number of studies have also attempted to correlate exposure to phyto-oestrogen-rich foods with clinical endpoints. In a series of comparative studies it has been demonstrated that significant hormonal changes to the menstrual cycle are observed when pre-menopausal women consume a diet containing 45 mg of isoflavones fed as soya foods.[7] Half of this dose was ineffective and physiological effects were not observed when an isoflavone-free soya product was fed.[8] It would appear that diets rich in soya may be oestrogenic. However, further clinical studies are necessary to test efficacy before isoflavone-containing foods can be considered to be 'functional' in nature.

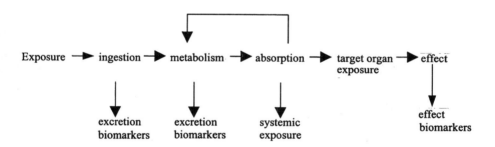

Figure 2 *The causality chain from exposure to effect*

The major problems associated with functional foods concern intake (exposure) and effect. In order to determine that a food has an effect beyond its nutritional attributes, it is necessary to confirm that the target population are exposed to the food which contains the putative active agent and that this particular component (and not some other dietary constituent) is having the desired effect. In order to prove this chain of causality leading from exposure to effect, it is necessary to have information on the mechanism by which an effect occurs (Figure 2). This mechanistic information can take many forms however, in general, the further along the chain of causality towards clinical outcome an event is measured, the more certain the correlation with the outcome is and the less sure the link to specific dietary exposure. The reverse is also the case. It has become widely accepted that the pre-clinical (and, by definition, pre-diagnostic) events are termed **biomarkers**.

It is clear from the above that, in order to determine the link between a dietary constituent and a health benefit, it is necessary to interrogate the chain of causality at a number of points by measuring biomarkers both of exposure and of effect. In the case of soya there are two populations for whom a causative pathway between exposure and effect may be demonstrable. The first of these are post (or peri) menopausal women. In this case, as noted above, there is some evidence to suggest that the isoflavones present in soya may have an distinct biological effect resulting in a number of clinical changes related to menopause. In addition, there is also some evidence of a positive effect on the risk of coronary heart disease through a reduction in total cholesterol and in low density lipoprotein. The clinical effects are late-stage health events and it is difficult to absolutely correlate these specifically with soya isoflavone intake. Since the diet of most adult women is complex and varied, the correlation between distinct effects and specific dietary constituents is difficult to confirm in free-living individuals.

At the other end of the scale of dietary complexity, a sub-group of neonates are fed soya-based formula. In this case the comparison can be easily made between soya and non-soya fed infants and the intake and absorption monitored. It is, however, considerably more difficult to determine any potential correlation between the highly defined diet and a clinical endpoint since the possible clinical effects are either not measurable in infants or are only manifest much later in life. In this case, the potential of soya formula as a functional food is again difficult to demonstrate conclusively.

In the present study, a comparison was made between infants fed on soya formula and those fed on cows' milk based formula or on maternal milk. By building up a picture of exposure profiles it should be possible to then develop further the correlation between exposure and effect through the use of appropriate biomarkers.

2 Experimental Design

Infants were recruited who were fed on either breast milk, soya or cows' milk formula (Table 1). Plasma and spot urine samples were taken and analysed for genistein, daidzein and equol, after hydrolysis to release any conjugates.

Table 1 *Subjects recruited fed on breast milk, soya or cows' milk formula*

Feeding regime	No. recruited	Age range (Weeks)	Samples collected		
			blood	urine	milk
Soya formula	14	12–60	11	11	
Cows' milk formula	15	6–51	11	8	
Breast fed	7	14–62	6	5	2

Blood samples were routinely analysed by gas chromatography/mass spectrometry (GC/MS). Deuterated (D_3) daidzein was used as an internal standard. Urine samples were analysed by HPLC/MS and the levels compared to matrix-matched standards. Because the urine analysis was carried out on spot samples (*i.e.* not 24-hour collections), the amounts of the isoflavones present were referenced to the concentration of creatinine.

3 Results

The results from the analysis of the blood and urine samples are given in Tables 2 and 3.

Table 2 *Levels of isoflavones found in spot urine samples. Figures given are means with the standard error given in brackets. An entry of – means none detected*

Diet	n	Levels ($\mu g/mg$ creatinine) Genistein	Daidzein	Equol
Soya	10	33.5 (13.7)	31.1 (12.4)	–
Breast fed	7	1.5 (0.7)	1.7 (0.9)	–
Cows' milk	8	0.08 (0.05)	2.2 (0.03)	0.005

Table 3 *Levels of isoflavones found in plasma. Figures given are means with the standard error given in brackets. An entry of – means none detected*

	Diet	n	Levels ($\mu g/mg$ creatinine) Genistein	Daidzein
Equol				
Soya	10	33.5 (13.7)	31.1 (12.4)	–
Breast fed	7	1.5 (0.7)	1.7 (0.9)	–
Cows' milk	8	0.08 (0.05)	2.2 (0.03	0.005

The limit of detection of genistein and daidzein in urine was 13 pg/μl and 11 pg/μl, respectively, based upon spiking with the unconjugated compounds and carrying out the hydrolysis followed by analysis. The limit of detection for both genistein and daidzein in serum was 2.5 pg/μl while that for equol was 12.5 pg/μl.

4 Discussion

These results demonstrate the exposure of neonates to isoflavones after feeding with soya. It is clear from the results that the compounds are absorbed and that metabolism to give equol does not occur. This may be a result of the lack of the appropriate gut microflora in the neonates. While it is not possible to determine the extent of the uptake, these results are broadly similar to results reported previously by Setchell *et al.*[9] in which levels of total isoflavones of the order of 25–50 ng/ml were found in urine from infants fed on soya formula. The results from the present study, when calculated in a similar fashion, suggest levels of up to 200 ng/ml; however, without referencing the results to creatinine, it is not possible to make valid comparisons. In the case of the analysis of the serum samples, Setchell reported a mean level of 979 ng/ml in soya-fed infants and, again, this is comparable to the levels found in this study (79–563 ng/ml).

Having demonstrated exposure and uptake of isoflavones in this very specific population with defined dietary characteristics, the next stage in the procedure is to attempt to travel further along the causality pathway and to demonstrate a potential effect that could be attributable to exposure to isoflavones at a young age. The most obvious target would be the induction of gene expression and the potential effect this could have at a developmentally important stage in life. By developing biomarkers of effect coupled to long term epidemiological studies it may be possible to delineate a beneficial effect arising from neonatal exposure to phyto-oestrogens.

References

1   K. D. R. Setchell and H. Adlercreutz, in 'Role of The Gut Flora in Toxicity and Cancer', ed I. R. Rowland, London, Academic Press, p. 315.

2   G. Leclerq and J. C. Hewson, *Biochim. Biophys. Acta*, 1979, **560**, 427.

3   G. G. J. M. Kuiper, E. Enmark, M. Peltohuikki, S. Nilsson, J-A. Gustaffson, *Proc. Natl. Acad. Sci* USA, 1996, **93**, 5925.

4   G. G. J. M. Kuiper, B. Carlsson, K. Grandien, E. Enmark, J. Haggblad, S. Nilsson and J. A. Gustaffson, *Endocrinology*, 1997, **138**, 863.

5   D. F. V. Lewis, M. G. Parker and R. J. B. King, *J. Steroid Biochem. Molec. Biology*, 1995, **52**, 55.

6   A. M. Brzozowski, A. C. W. Pike, Z. Dauter, R. E. Hubbard, T. Bonn, O. Engstrom, L. Ohman, G. L. Greene, J.A. Gustafsson and M. Carlquist, *Nature*, 1997, **389**, 753.

7   A. Cassidy, S. Bingham and K. D. R. Setchell, *Amer. J. Clin. Nutr.*, 1994, **60**, 333.

8   A. Cassidy, S. Bingham and K. D. R. Setchell, *British J. Nutrition*, 1995, **74**, 58

9   K. D. R. Setchell, L. Nechemias-Zimmer, J. Cai and J. E. Heubi, *Lancet*, 1997, **350**, 23.

Phytomedicine for Thermogenic Stimulation in Obesity Management: Potentials and Limitations

A.G. Dulloo

INSTITUTE OF PHYSIOLOGY, UNIVERSITY OF FRIBOURG, RUE DU MUSÉE 5, CH-1700 FRIBOURG, SWITZERLAND

1 Introduction

In this era when information technology is revolutionizing health awareness among the general public, more and more people are increasingly conscious of the health hazards of becoming fat. This change in perception of excess adiposity from a 'cosmetic' to a 'health' issue is encouraging since it is now acknowledged by the medical establishment that even a modest degree of obesity increases the risks for chronic diseases traditionally associated with more severe obesity, namely type 2 diabetes, cardiovascular diseases, and certain forms of cancer. However, the cornerstone methods for managing body weight – by dietary restriction and/or by exercise – have proven to be largely ineffective since few people can stick to the dietary regime or exercise therapy. The result is generally a transient phase of weight loss (or weight stability) followed by a return, within a few years, on the trajectory towards obesity. These failures to prevent and treat obesity have led to a re-examination of classical views about the causes of this disorder, and indeed a reconsideration of the general concepts of homeostatic mechanisms that regulate body weight. Although the increasing prevalence of obesity is associated with excess food/fat intake and reduced physical activity of a modern life, there is also growing realization that genetic susceptibilities for 'slow metabolism' (e.g. low basal metabolic rate, low capacity to oxidize fat) also play an important role in determining the extent to which an individual resists or is prone to excess fat deposition. Furthermore, it is now recognized that in response to reduced food intake (by dieting or with the help of anorectic agents), there is an accompanying reduction in energy expenditure which is in part due to loss of body mass, and in part due to a slowing-down in the metabolic activity of the remaining tissues. It is towards these conditions of 'slow metabolism', characterized as diminished thermogenesis, that considerable research *vis-a-vis* obesity therapy is directed: that is to say, there is an active search for means of increasing thermogenesis and fat oxidation in weight management.

2 From Thyroid Extracts to Functional Foods

The idea of stimulating thermogenesis to treat, or to assist the treatment of obesity has a long history. Thyroid extracts (subsequently thyroid hormones) and uncoupling agents (like dinitrophenol) were utilized in thousands of obese patients earlier this century. Although they induced marked reductions in body weight, their use in obesity therapy fell into disrepute because of unacceptable side-effects. Numerous other compounds, which can be categorized as hormones (*e.g.* glucagon, growth hormone), synthetics (*e.g.* dinitrophenol-derivatives, vasodilators such as nicotinic acid, salicylates), or foods (*e.g.* certain amino-acids like glycine, citrus extract, liebig extract) were subsequently tested for thermogenic properties, but their effects on metabolic rate were found to be either insignificant or of too short duration.[1] A turning-point in the screening for thermogenic agents occurred in the early 1980s when a role for the sympathetic nervous system (SNS), *via* its heat-producing neurotransmitter noradrenaline (NA), became implicated in the dietary regulation of thermogenesis. The demonstrations that reduced SNS activity could play a role in the adaptive reduction in metabolic rate that is counter-effective to dieting and that low level of SNS activity may underlie the 'slow metabolism' that predispose certain individuals towards obesity, provided the physiological rationale for mimicking the effect of NA in stimulating thermogenesis and fat oxidation in obesity therapy.[2] The pharmaceutical approach has concentrated on the development of drugs that would target atypical (β_3) adrenoceptor, believed to be the pivotal receptor *via* which sympathetically-released NA activates thermogenesis in various tissues.[3] In parallel, there has also been considerable interest in the nutritional/nutraceutical areas for screening foods and dietary ingredients with potential thermogenic properties by virtue of their mode(s) of action by interference with the SNS.[4] There are several key control points for such interference along the line of sympathetic control of thermogenesis, and these include (i) enhancing NA release, *(ii)* activating the membrane bound adenylate-cyclase system, *via* which interaction between NA and its receptors leads to intracellular formation of cAMP (the second messenger that drives cellular thermogenesis), *(iii)* blocking inhibitory modulators of NA release, and action in the synaptic cleft (*e.g.* adenosine and certain prostaglandins), and (iv) inhibiting phosphodiesterases, the enzyme system that degrades NA-induced cAMP in the cytoplasm. Against this background, this paper reviews briefly the various plant extracts or ingredients that have generated the most interest, and for which there is some evidence in humans concerning their potentials as safe and efficacious thermogenic agents for weight control. These include phyto-products such as ephedrine-rich ephedra herbs, caffeine-rich coffee, catechin-rich green tea, capsaicin-rich red pepper, and medium-chain triglycerides (MCT) rich coconut oil.

2.1 Ephedra Herbs and Ephedrine

L-ephedrine is an alkaloid with sympathomimetic properties that is found in several species of *Ephedra*. In the form of the herb Ma Huang, it has been used for ritualistic and medicinal purposes in China more than 2000 years before it was introduced in Western medicine in the 1920s. By virtue of its property as a long-acting catecholamine (ephedrine's main mode of action is to enhance the release of NA from sympathetic nerves), it found utility for a variety of therapeutic use, and it is still found nowadays in many preparations (some over-the-counter) as a treatment for asthma, bronchitis and nasal congestion. The stimulatory effect of ephedrine on energy metabolism was first brought to attention in 1925 by Chen and Schmidt[5] who pointed out that 100–125 mg of ephedrine increased the resting metabolic rate and fat oxidation in humans. However, half a century elapsed before this property of ephedrine was revived by Evans and Miller[6] who advocated clinical trials with this sympathomimetic following the demonstration that acute oral administration of ephedrine (40 mg) stimulated thermogenesis in humans, and was effective in reducing fat in several animal models of obesity. A number of clinical trials in obese patients on hypocaloric diets have since shown the efficacy of ephedrine to induce greater weight loss than a placebo.[7] While these anti-obesity effects are often attributed to its appetite-suppressing properties (hence enabling better compliance to dietary regimes), the contribution of its thermogenic properties to weight loss is now recognized.[8] Several herbal products based on Ma Huang (*Ephedra sinica*) or capsules containing various *Ephedra* herbs are commercially available but, to date, there are no scientific data on the efficacy of these herbal products in stimulating thermogenesis. Furthermore, these products contain not only L-ephedrine, but also several isomers of ephedrine such as norephedrine, pseudoephedrine and norpseudoephedrine, for which data about efficacy and safety as anti-obesity thermogenic agents are also absent.

2.2 Coffee Beans and Caffeine

The ability of caffeine (the most abundant dietary methylxanthine) to stimulate metabolic rate in humans was first demonstrated more than 80 years ago. The attribution of the thermogenic effect of coffee solely to its content in caffeine is now well established,[9] and of particular interest since caffeine is known to be capable of interfering at several control points along the SNS-thermogenesis axis, in particular by inhibiting phosphodiesterases, thereby potentiating NA-induced thermogenesis.[10] A meta-analysis of literature data on the effect of single-dose administration of caffeine on resting metabolic rate after an overnight fast reveals a linear relationship (unpublished findings), with the threshold for a detectable stimulatory effect (5% increase) occurring between

80 and 100 mg and a 40% increase above resting metabolic rate obtained with one gram of caffeine. Under confined conditions of a respiratory chamber, repeated administration of 100 mg caffeine (equivalent to 1–2 cups of coffee) at regular intervals between 8:00 hour and 18:00 hour (*i.e.* total of 600 mg/day) was found to increase 24 h energy expenditure by 5% in both lean and post-obese individuals.[11] More recently, repeated intake of coffee containing higher doses of caffeine (250–300 mg, 5 times/day) was found to increase 24 h energy expenditure by 8% and 5% in lean and obese subjects, respectively, without apparent side-effects.[12] These thermogenic effects of caffeine/coffee are shown under weight maintenance conditions but, during hypocaloric treatment, obese individuals ingesting 200 mg three times a day (600 mg/day) lost no more weight than those on placebo.[13] Overall, coffee/caffeine certainly has thermogenic effects which might assist in the prevention of obesity, but these effects would seem to be absent under conditions of hypocaloric intake, possibly because of the accompanying reduction in SNS activity, and hence less phosphodiesterase activity to inhibit. This contention is supported by the fact that low doses of methylxanthines like caffeine were found to be effective in potentiating the effect of ephedrine (an enhancer of NA release) on thermogenesis,[14] and that a combination of ephedrine and caffeine resulted in greater fat losses that either ephedrine or caffeine alone, in women on a hypocaloric diet; effects which were shown to be well-tolerated, with side-effects being few and only transient.[13, 15]

2.3 *Coleus forskohlii* Roots and Forskolin

The roots of *Coleus forskohlii* is commonly used as folk medicine in Asia for various ailments including abdominal colic, respiratory disorders, heart diseases, insomnia and convulsions. These effects are generally thought be due to forskolin, which is the major diterpene isolated from the plant and whose pharmacological properties (best illustrated by its positive inotropic action and potent vasodilatory properties) are attributed to the activation of the membrane-bound enzyme, adenylate cyclase. Since the net driving force for thermogenic stimulation under SNS control resides in sustained elevation of cellular levels of cAMP resulting from β_3-adrenoceptor activation of the adenylate cyclase enzyme, this enzyme system itself provides an additional target for pharmacological interference aimed at enhancing thermogenesis. Indeed, forskolin is well known to be a potent stimulator of adenylate cyclase, cAMP formation and lipolysis *in vitro*.[16] In a recent experiment using our room calorimeter to investigate the efficacy of a single dose oral administration of an extract of *coleus forskohlii* containing 10–20 mg of forskolin, *i.e.* doses previously reported to be well tolerated,[17] we found a 3–4% increase in the resting metabolic rate of lean and obese subjects (unpublished data). At such doses, this thermogenic effect of *coleus forskohlii* appears to be small, but the fact that it is sustained for about 6 hours, with no change in heart rate, blood pressure or protein oxidation and uncompensated during the remaining part of the 24 h cycle, raises the possibility that its repeated administration (2–3 times) daily may be effective in stimulating thermogenesis to the extent of increasing 24 h energy expenditure, without cardiovascular or other side-effects.

2.4 Coconut Oil and MCT

Fat in the typical western diet consists essentially of long-chain triglycerides (LCT) which are derived from animal fat and vegetable oils, with a minor component as medium-chain triglycerides (MCT) which are derived from dairy products. While LCT yield fatty acids that have a chain length of 14 carbons or more (usually C14–C22), MCT yield fatty acids that have a carbon length of 12 carbons or less (usually C6–C12). The observation in clinical medicine that during nutritional rehabilitation of cachexic patients, administration of lipid emulsions rich in MCT are more readily oxidized than LCT, has generated considerable interest in investigating the role of MCT in the prevention or treatment of obesity.[18] Several single-meal response studies have shown that MCT tends to increase post-prandial thermogenesis to a greater extent than LCT in both lean and obese individuals. Although this greater thermogenic effect of MCT is generally attributed to the higher energy cost of its metabolic fate, there is also evidence that its administration can lead to central activation of sympathetic activity. In a dose–response study[19] investigating the thermogenic potential of MCT when consumed as an integral part of the typical western diet, we have shown that 5–10g of MCT ingested with each of the three meals (breakfast, lunch and dinner), in substitution of LCT, stimulated thermogenesis to the extent that 24 h energy expenditure was increased by 5%. Furthermore, these effects on metabolic rate were accompanied by a significant increase in urinary noradrenaline and not adrenaline,[19] and hence in support of the notion that MCT-induced thermogenesis may be mediated *via* a central activation of sympathetic outflow. There have also been a few studies investigating the impact of feeding (for 1–2 weeks) diets rich in MCT on energy balance and body composition in humans.[20] However, the results are somewhat inconclusive about whether the greater thermogenic response to MCT than LCT is sustained and whether BMR is elevated during chronic administration. One possible explanation for this controversy may reside in the composition of MCT. In some of the studies, an increase in MCT was achieved by increasing dairy product consumption, *i.e.* MCT which yields predominantly C12 fatty acids, and it is possible that it is MCT with a high proportion of C6–C10 fatty acids that confers the greatest potency in stimulating thermogenesis. The potential use of such MCT high in C6–C10 (usually extracted from coconut oil) as the lipid source in various food items and confectionery is currently the subject of investigation but there is clearly a need for clarification about the type of MCT than confers thermogenic potential. Furthermore, since MCT are invariably saturated, their impact on weight control needs to be studied concomitantly with their effects on blood cholesterol and other indicators of cardiovascular diseases.

2.5 Spices, Red Pepper and Capsaicinoids

In the search for dietary components that would promote fat oxidation and thermogenesis, the pungent spices in foods have also attracted interest,[21] triggered initially, perhaps, by their apparent ability to 'subjectively' warm the body. More objective evidence for spice-induced thermogenesis was first provided by Henry and Emery[22] who showed that chilli and mustard sauces augmented the thermogenic response to a meal. More recently Yoshioka *et al.*[23] have reported that red pepper enhanced thermogenesis and lipid oxidation in women. It is now established from animal studies that it is the principle ingredients that confer pungency to these spices and to others such as Tabasco sauce and ginger, namely the capsaicinoids (capsaicin and dihydrocapsaicin) and capsaicinoid-homologues (the gingerols and shogoals), that also confer the thermogenic and lipolytic properties. Of particular interest in the red pepper study[23] is that these metabolic effects of the capsaicin-rich foods were demonstrated in women showing diminished meal-induced thermogenesis and failure to increase fat oxidation when shifted from a high-carbohydrate to a high-fat meal. The addition of the capsaicin-rich red pepper to the high-fat meal was found to stimulate fat oxidation, and to normalize the thermogenic response to the high-fat meal to levels found with the high-carbohydrate meal. In other words, spices rich in capsaicinoids have the potential for adjusting fat oxidation in response to fat intake. These metabolic effects of pungent principles of spices are perhaps not unexpected in view of evidence suggesting that their main modes of action can also be linked to interference with the sympathoadrenal system. By exerting their metabolic effects either centrally to induce adrenal medullary secretion or peripherally by interference with sympathetic control in tissues such as brown adipose tissue, spices rich in capsaicinoids could therefore constitute a new class of dietary ingredients with sympathomimetic thermogenic effects.[21]

2.6 Green Tea and Catechin-Polyphenols

Various extracts of green tea, a widely consumed beverage in China and Japan, are frequently claimed to be effective in reducing body weight. These anti-obesity properties, although unsubstantiated, are generally attributed to their caffeine content. But green tea is also rich in polyphenols, some of which, namely the flavanols (*e.g.* catechin-polyphenols) and flavonols (quercetin, myricetin) have been shown to inhibit catechol-*o*-methyl transferase (COMT), the enzyme that degrades the catecholamines (noradrenaline and adrenaline). Thus, by virtue of its high content both in catechin-polyphenols (which may hence reduce the degradation of noradrenaline within the synaptic cleft) and caffeine (which inhibits phosphodiesterase within the cytoplasm), green tea has thus the potential to interact synergistically with the SNS leading to the potentiation and prolongation of noradrenaline-induced thermogenesis. Such a hypothesis has been shown to hold true when tested in brown adipose tissue *in vitro.*[24] In a subsequent placebo-controlled study in lean and overweight men, we studied the effect of ingesting capsules containing a green tea extract (AR25), the dry weight of which comprises as much as 25% catechins and 7% caffeine, on 24 h energy expenditure and substrate oxidation in a room calorimeter.[25] This green tea extract was found to stimulate thermogenesis to the extent of increasing 24 h energy expenditure by about 4%, and to promote fat oxidation such that the contribution of fat oxidation to daily energy expenditure was increased from 30% to 40%. These effects cannot be explained by its caffeine content *per se*, since administration of the same amount of caffeine as in the green tea extract failed to increase 24 h energy expenditure or fat oxidation.[25] These data *in vitro* and *in vivo*, together with the findings in the human study that 24 h urinary noradrenaline (and not adrenaline) was higher during treatment with this green tea extract than with placebo or caffeine alone, therefore support the notion that the efficacy of the green tea extract in stimulating thermogenesis and in promoting fat oxidation is likely to reside in an interaction between catechin-polyphenols, caffeine and sympathetic activity. The extent to which the therapeutic value of green tea, which is believed to protect against chronic diseases by virtue of its potent anti-oxidant properties, can be extended to the management of obesity will have to await the results of a planned placebo-controlled clinical trial with this green tea extract in obese humans.

3 Concluding Remarks

Phytomedicine is increasingly advocated for weight control, as judged by numerous articles that appear in the popular press and magazines. Among the most popular are those under the category of 'fat burners' but, to date, there is little, if any, robust scientific basis to support such claims for the vast majority of phytoproducts that are being proposed to consumers. However, phytotherapy is increasingly appealing to a large section of the public, and although this topic in obesity research is only in its infancy, the *potential* for certain plant extracts or plant ingredients to exert quantitatively important stimulatory effects on thermogenesis and fat oxidation is real, and those discussed above are well within the range of thermogenic stimulation so far obtained with more 'selective' drugs. However, to gain credibility and acceptance by the mainstream research community and by the medical establishment, there is a need for industry to commit itself to a more systematic screening of phytoproducts for thermogenic efficacy, to better understand the way their (often more than one) active ingredients interact with the biological system(s), and to test the safety and efficacy of their candidate anti-obesity phytoproducts with the same scientific rigour as that required in the development of novel anti-obesity drugs. Considering the large number of drugs in clinical use today whose origin can be traced to 'medicinal' plants, the chances are that the search for thermogenic 'phytoproducts' will eventually lead to the discovery of new classes of thermogenic drugs that would be of value to assist in the management of obesity and its debilitating metabolic complications.

References

1   A.G. Dulloo and D.S. Miller, *Wrld. Rev. Nutr. Diet*, 1987, **50**, 1.
2   L. Landsberg and J.B. Young, *Int. J. Obes.*, 1992, **17** (Suppl. 1), S29.
3   M.J. Stock, *Nutrition in Prevention of Diseases* (J. Somogyi), Basel: Karger, 1989, p. 32.
4   A.G. Dulloo, *Nutrition*, 1993, **9**, 366.
5   K.K. Chen and C.F. Schmidt, *J. Pharmacol. Expt. Ther.*, 1925, **24**, 339.
6   E. Evans and D.S. Miller, *Proc. Nutr. Soc.*, **36**, 136A.
7   R. Pasquali and F. Casimirri, *Int. J. Obes.*, 1992, **17** (Suppl. 1), S65.
8   A.G. Dulloo and D.S. Stock, *Int. J. Obes.*, 1992, **17** (Suppl. 1), S1.
9   K.J. Acheson, B. Zahorska-Markiewicz, Ph. Pittet, K. Anantharaman and E. Jéquier, *Am. J. Clin. Nutr.* 1980, **33**, 989.
10 A.G. Dulloo, J. Seydoux and L. Girardier, *Metabolism* 1992, **41**, 1233.
11 A.G. Dulloo, C. Geissler, T. Horton, A. Collins and D.S. Miller, *Am. J. Clin. Nutr.* 1989, **49**, 44.
12 D. Bracco, J.M. Ferrarra, M.J. Arnaud, E. Jéquier and Y. Schutz, *Am. J. Physiol.*, 1995, **269**, E671.
13 S. Toubro, A. Astrup, L. Breum and F. Quaade, *Int. J. Obes.*, 1992, **17** (Suppl. 1), S69.
14 A.G. Dulloo, *Int. J. Obes.*, 1992, **17** (Suppl. 1), S35.
15 P. Daly, D. Krieger, A.G. Dulloo, J.B. Young and L. Landsberg, *Int. J. Obes.*, 1992, **17** (Suppl. 1), S73.
16 B.S. Kenneth, *Ann. Rep. Med. Chem.*, 1984, **19**, 293.
17 P. Meyer, 'Report of a clinical study', *Hôpital Necker*, Paris, 1993.
18 A.C. Bach and V.K. Babayan, *Am. J. Clin. Nutr.*, 1982, **36**, 950.
19 A.G. Dulloo, M. Fathi, N. Mensi and L. Girardier, *Eur. J. Clin. Nutr.*, 1996, **50**, 152.
20 A.A. Papamandjaris, D.E. MacDougall and P.J.H. Jones, *Life Sciences*, 1998, **62**, 1203.
21 A.G. Dulloo, *Br. J. Nutr.*, 1998, **80**, 493.
22 C.J.K. Henry and P. Emery, *Hum. Nutr. Clin. Nutr.*, 1986, **40C**, 165.
23 M. Yoshioka, S. St-Pierre, M. Suzuki and A. Tremblay, *Br. J. Nutr.*, 1998, **80**, 503.
24 A.G. Dulloo, J. Seydoux and L. Girardier, *Int. J. Obes.*, 1996, **20** (Suppl. 4), 71A.
25 A.G. Dulloo, C. Duret, D. Rohrer, L. Girardier, N. Mensi, M. Fathi, P. Chantre and J. Vandermander. *Am. J. Clin. Nutr.*, 1999, in press.

4 Micronutrients

Folic Acid – A Case Study for Fortification

J.M. Scott

BIOCHEMISTRY DEPARTMENT, TRINITY COLLEGE,
DUBLIN 2, IRELAND

1 Introduction

This paper summarises the now incontrovertible evidence that folic acid or folate have a role in the prevention of the congenital conditions collectively referred to as neural tube defects (NTDs). It is now clear that this effect is not due to treatment of maternal folate deficiency at least in the conventional sense. For reasons dealt with later it is clear that the effective use of folic acid to prevent NTDs will be achieved by fortification of the diet with folic acid. What is emerging is that genetic variants of folate dependent processes, perhaps quite common in occurrence, produce an increased risk of incomplete closure of the neural plate in certain embryos. These embryos will inherit this genetic make up from both parents. It is also emerging that folate status over the physiological range has an influence on whether a NTD occurs in a susceptible embryo. Embryo folate status in turn depends upon maternal folate status and thus the mother fulfils two roles – one in passing on half of the genetic information to the embryo and the other in determining the embryo's folate status.

It is clear that prevention of NTDs, which should have ensued from this knowledge, has not happened. This is because the neural tube closes before most women recognise that they are pregnant and consequently extra folic acid/folate, to be effective, must be taken before the outcome of the pregnancy is known. Thus women need to take folic acid periconceptionally (*i.e.* before conception and during early embryonic development). In most countries it is known that over half of pregnancies are unplanned (perhaps not unwanted when they happen, but nevertheless unplanned). It is clear that women will not make elective changes in their diet, either by changing the pattern of the foods they eat, to take more folate rich foods, to take fortified foods or to take supplements, to prevent an event in a pregnancy that they don't plan on having. Thus over half of women will not benefit. Even for the half who do plan their pregnancy, it has proven very difficult despite extensive education, firstly to get across the benefit of taking folic acid in preventing NTDs and, secondly, even in women who are aware of the benefit, to get them to take the required extra folic acid on a daily basis. This, it will be seen, produces a situation where the only likely effective measure is either mandatory or extensive voluntary fortification of the diet with folic acid.

2 The Functions of Folate

The role of folate in late pregnancy, in preventing the megaloblastic anaemia of pregnancy, was first identified by Lucy Will working with pregnant women in India some sixty years ago, and is a different phenomenon.[1] This anaemia was due to the increased demands that arise from the rapidly growing placental/foetal structures.[2] It has been shown that this rapid and extensive cellular development accelerates the catabolism (breakdown) of folate.[3] This rate of folate catabolism cannot be matched by a normal diet and the negative folate balance then ensues, producing maternal folate deficiency and anaemia in women who have not entered pregnancy with adequate stores or who do not receive prophylaxis with folic acid. This role of folate in pregnancy has been established and acted upon in most instances for decades. The role of folic acid in preventing NTDs is a newer and completely different phenomenon. The megaloblastic anaemia of pregnancy relates to the essential role that folate in cells plays in cell division.[4] The folate cofactors, as the reduced form tetrahydrofolate, are involved in the transfer of so-called carbon one groups found in cells as 10-formyltetrahydrofolate and 5,10-methylene tetrahydrofolate (Figure 1). These cofactors pass on a formyl group to enzymes in the purine biosynthase pathway and as such are the origins of carbon 2 and carbon 8 of the purine ring (Figure 2). The methylene group is passed on to the enzyme thymidylate synthase to convert the uracil-type base found in RNA to the thymine-type base found in DNA. As such, folates are essential for purine and pyrimidine biosynthesis and in turn thus needed for DNA biosynthesis. Reduction in folate status will limit such synthesis and reduce the rate of cell division in all cells. This will be most apparent in rapidly dividing cells such as those of the bone marrow and most easily seen as anaemia. Thus reduced folate status has long been associated with anaemia. The role of folic acid/folate in preventing NTDs is not due to such gross deficiency as would cause anaemia. The size of the embryo at the time of closure of the neural tube is about the size of the head of a pin and so the embryo has not significantly drawn on maternal stores of folate. Thus embryonic demands at this stage do not compromise maternal folate status. However, the closure of the neural tube is clearly responsive to embryonic folate status during the critical four days of closure (days 21 to 27). This status is in turn dictated by maternal folate status before closure commences. The early nature of this event has become critical in implementing a program to prevent folate related NTDs. Most pregnancies are unplanned and thus, even if they know of the potential benefits, many women will not themselves be taking folic acid supplements.

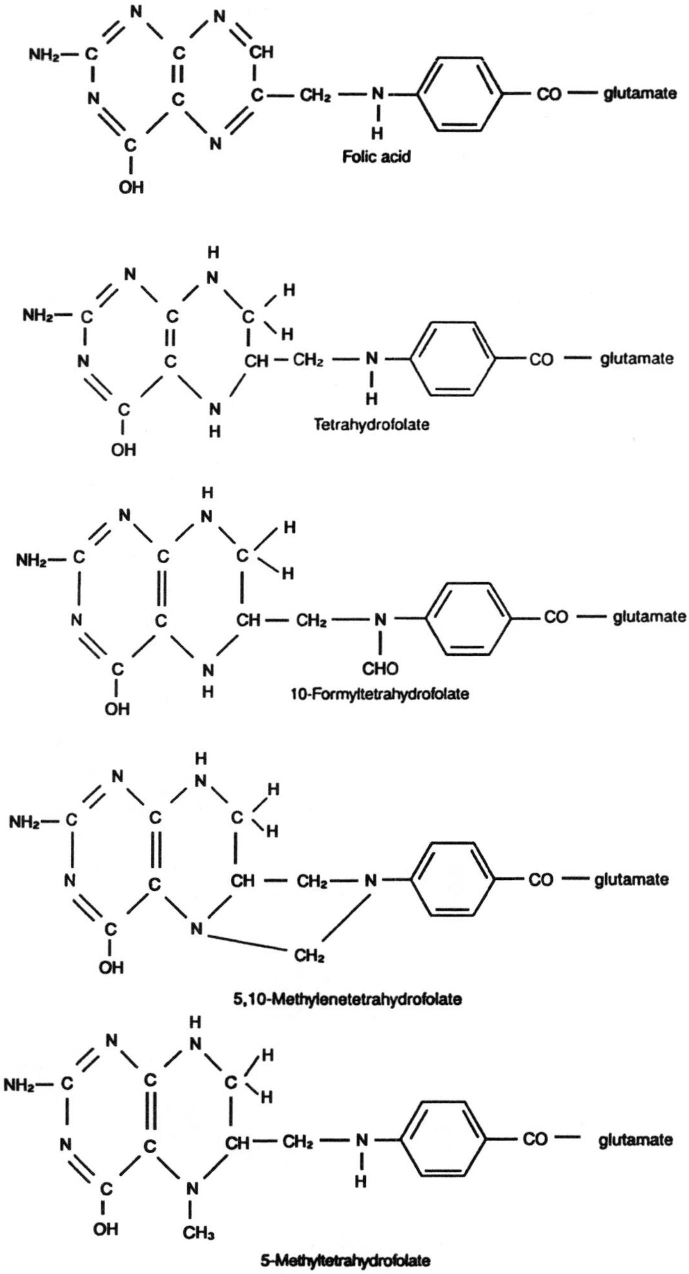

Figure 1 *The formula of synthetic form of the vitamin folic acid and of the most important naturally occurring folates*

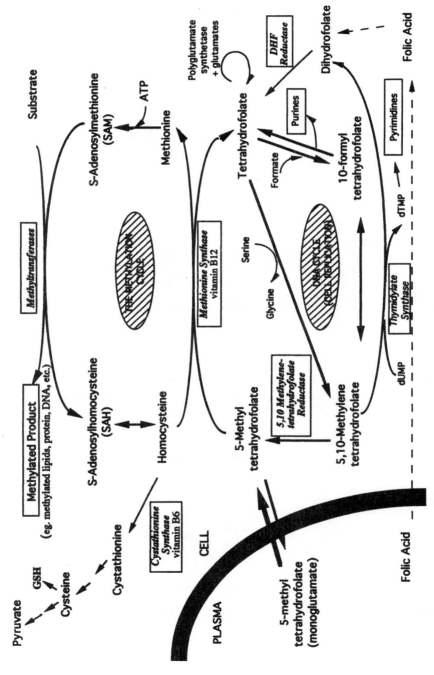

Figure 2 *The role of the folate cofactors in the biosynthesis of purines and pyrimidines and thus DNA and in the methylation cycle.*

Equally, by the time the neural tube has closed at day 27, the majority of women would not have realised that they are in fact pregnant, precluding an effective response at that stage. Thus despite the fact that there is total agreement that maternal periconceptional taking of folic acid can prevent the majority of NTDs, there has been an almost total failure in implementing such prophylaxis.

3 Definition of Neural Tube Defects (NTDS)

Early in embryonic development, the neural plate folds to form the structure of the neural tube, which will eventually encompass the spinal column.[5] A protrusion of this structure will form the cranium which will ultimately enclose the brain. Incomplete closure of the former leads to a spinal column with an opening in it, leading to the clinical condition called spina bifida. The severity of the condition depends largely on the size and position of the opening and can vary from being undetectable clinically, to where the child has paralysis of the bladder and the lower limbs and hydrocephalus due to interruption of the flow of the cerebrospinal fluid from the brain. Incomplete closure of the cranium results in what is clinically called anencephaly and is incompatible with life, with the child dying either *in utero* or within hours or days after birth. Spina bifida and anencephaly represent 50% and 40% respectively of what are collectively called neural tube defects (NTDs) with a variety of other variations contributing the remaining 10%. There is good epidemiological evidence, principally based on the way that either condition can recur in a susceptible mother, to suggest that NTDs have the same aetiology.[6]

4 Factors That Determine the Prevalence of NTDS

A great deal of interest has existed over the years in the factors that determine the prevalence at birth of NTDs. This interest has been driven by the hope that identification of factors that determine prevalence would be helpful in determining the aetiology of the condition, eventually leading to its prevention. It has emerged that the main determinants of prevalence are a genetic predisposition interacting with an environmental factor or factors.

4.1 Genetic Predisposition

It has been recognised for decades that there is a genetic component in the aetiology of NTDs.[6] The strongest evidence implicating a genetic component is that having a previous pregnancy affected by an NTD, increases risk of what is called a recurrence by some 10 to 15 times. After two or more such births the prevalence gets as high as 10% risk of recurrence. Also risks to close (first degree) relatives of such women have been established as being higher than in the general population. Further support for a genetic component is that concordance rates are higher in monozygotic than dizygotic twins.[6] There are well established racial differences, rates are usually low in African-Americans, while traditionally they have been high in populations of Celtic origin. Even when such populations migrate, prevalence may remain high, as in Western Australia.[7,8]

There may be common environmental factors in families and in ethnic groups. However, the high increased risk of recurrence where the prevalence changes so dramatically would be hard to explain on the basis of some environmental factor that would not be obvious.

While there is thus an established polygenic predisposition for a given genetic background, one must keep this risk in balance. The risk is higher – perhaps ten-fold higher – but only in relative terms. Women with previously affected pregnancies would go from a risk – say in the general community – of 2 per 1000 births to 2 per 100 births. This absolute risk is still only 2%. Put another way, such a mother could, on the vast majority of occasions, be expected to have a normal pregnancy. Part of the explanation for this low prevalence may be due to the known fact that many fetuses with NTDs are aborted during early embryonic life. In fact, prevalence of NTDs should properly be referred to prevalence at birth, since it is this number that is used. In addition, it seems likely that the genetic components that partly determine prevalence are those of the embryo. Thus the paternal genetic inheritance will also cause a variation in prevalence.

4.2 Environmental Predisposition

If variation in genetic background was the only factor influencing risk of an NTD affected birth, then the prevalence in a population where the genetic pool was not varying would be constant. There has been ample evidence that this is not the case. Variations have been reported over time, with increases seen frequently after periods of nutritional deprivation most notably following the Depression in the US in the 1920s and the post-war Dutch Famine in the late 1940s.[6] Seasonal variation in prevalence has also been reported with suggestions that prevalence is highest in children conceived during periods when fresh food is least available.[6] There is also increased prevalence in births to women of lower socioeconomic circumstances. The above would point to reduced intake of a nutrient or nutrients as being an important cause of NTDs.

The earliest evidence implicating folate/folic acid in the aetiology of NTDs was a series of studies implicating antifolates, principally aminopterin, as causing NTDs.[4] These observations were after exposure of humans and of animals to antifolates. The former were really just uncontrolled case reports while the latter animal studies were not really extensive or systematic. In addition, both in early studies and more recently in a very well controlled study, it was found that folate deficiency does not increase the prevalence of NTDs in rodents.[9] In the early 1960s, Hibbard and his colleagues suggested that folate deficiency, as well as causing maternal anaemia, might also be a factor in abruptio placenta and perhaps birth defects. Studies by Hibbard[10] and other groups examined relatively small numbers and results were contradictory, with some supporting such a hypothesis[11] and others producing evidence against it.[12] In retrospect it would have been unlikely because of the small numbers involved that these clinical studies would have produced clear-cut evidence that folate or nutrition were involved in the aetiology of NTDs. It was Smithells and his colleagues[13] who were the first to report a systematic effort to test the role of nutrition in the aetiology of NTDs. Smithells had in his earlier career worked with Hibbard. Not being clear which nutrient to examine, he tested a cocktail of nutrients in a commercial preparation called Pregnivite Forte F. This preparation contained five trace elements and six vitamins, including folic acid. The regimen decided upon was ingestion of tablets three times per day, giving a total daily dose of 360 µg of folic acid per day. It was Smithells's original intention to use a randomised placebo-controlled design. However, one of the three hospital ethics committees that he submitted his proposal to insisted that all women in the trial received some treatment. Smithells has been criticised retrospectively for accepting this decision and changing his design to an intervention study and he was also criticised for using a cocktail of nutrients. This decision had serious consequences. When the trial of Smithells *et al.*[13] demonstrated an apparently significant protective effect of the preparation when compared to the prevalence in women from the same regions, who, for one reason or another, had not entered the trial, the majority of medical opinion would not accept the result as being conclusive. Even when Smithells *et al.*[14] added further numbers, most experts in this field argued that there might be an intrinsic difference between those who entered this trial and those who did not, and that this explained the perceived benefit. Had the trial been properly controlled, the role of nutrients in the protection from NTD affected births might have been established a decade earlier. The historic, if flawed, studies of Smithells and his colleagues did, however, add a great incentive for others to investigate if nutrients or folic acid had a role in prevention of NTDs.

After Smithells *et al.* published their first results in 1980 Laurence *et al.*[15] examined the results of a double-blind randomised trial comparing folic acid at 4.0 mg per day with placebo to prevent recurrence of NTD. The initial results had been negative because they had observed two NTD affected births in the folic acid treated group. However, they felt justified retrospectively in considering both of these women as being 'non-compliers' (one on the basis of an admission and the other because her serum folate was the lowest in the treatment group). This re-analysis showed a protective effect for folic acid but the result was not generally accepted, because of the transfer of the two women with affected pregnancies into the non treated group. In addition, the trial involved very small numbers.

The study that established the role of folic acid in the prevention of recurrence of NTDs was undoubtedly that conducted by Professor Nick Wald under the auspices of the UK MRC.[16] This was a multicentre double-blind trial with four arms: folic acid, folic acid plus multivitamin, multivitamin alone and placebo. The two arms that contained folic acid were protective to the level of 72%, when compared to the multivitamin and placebo groups. The trial was of excellent design, with good compliance and large numbers and was accepted as giving conclusive proof that folic acid taken periconceptionally could prevent nearly three quarters of NTD affected births. There followed other trials, for example a trial in Southern Ireland which, while well designed, was too small to be conclusive.[17] An important advance was the trial of Czeizal and Dudas[18] which compared a multivitamin preparation with a multimineral preparation in women with no history of a previous NTD affected pregnancy, *i.e.* testing the effectiveness on preventing occurrence. This trial found that the former preparation, which contained folic acid, prevented the occurrence of NTDs in those in that arm. Because the trial used women with no previous history of an NTD affected birth, who have a much lower prevalence than the women with a previous history and in recurrence trials of Smithells and Wald, the number of embryos in the untreated group was only six. Thus while significant and appearing to give total prevention, the numbers involved were small.

Simultaneous with most of these latter trials, a series of case control studies emerged which showed retrospectively that in women who happened to be taking vitamin supplements that contained folic acid there was a definite protective effect, usually of the order of over 50%. As reviewed by Wald,[5] in all, there have now been seven case control studies, all but one of which showed a protective effect of folic acid.

While the conclusive MRC trial had used 4.0 mg per day it was generally felt that taking Smithells studies, which most people now regard as positive, and the case control studies that showed the standard vitamin preparation of 400 μg folic acid per day also to be protective, that this lower dose was probably adequate. This seemed reasonable since even this dose represents twice the RDA in most countries.

The UK Dept. of Health[19] and the CDC[20] in the US recommend that women who plan a pregnancy should take 400 μg of folic acid per day periconceptionally to prevent the occurrence of an NTD affected birth. The amount recommended for women who have had a previously affected birth is 4.0 mg/d to prevent recurrence.

5 What is the Mechanism by which Folic Acid Prevents NTDS

Essentially the following options appear to exist: that the prophylactic folic acid was (a) treating deficiency, (b) treating an intestinal malabsorption in susceptible women, (c) correcting a folate-dependent enzyme or process, or a combination of these mechanisms.

5.1 Maternal Folate Deficiency as the cause of NTDs

At first sight it would have appeared relatively simple to determine if women who had NTD affected pregnancies were folate deficient. In addition, if the aetiology was maternal malabsorption, it would have its effect through low folate status. The problem is that it was usually unclear which births were affected and which were not until, at the minimum, very late in pregnancy and usually post partum. This would have meant determining folate status of women at a point in pregnancy when their folate metabolism is known to be profoundly altered. Attempts were made to overcome this, in two studies, by using blood serum samples collected early in pregnancy for other purposes.[21,22] Both of these studies found no evidence of deficient serum folate levels in women who were found to have an NTD affected child in this, the index pregnancy. However, the numbers involved were small and it is recognised that serum folate is a poor index of status.

A prospective study was undertaken by us to collect samples to analyse red cell folate as well as serum folate levels.[17] Such samples were taken at the first visit of the pregnant woman to hospital and were then usually a few weeks or even months after the closure of the neural tube. However, it is recognised that red cell folate is a very reliable index of average folate nutrition. This is because during the five or so days during development of the red cell in the bone marrow, folate is incorporated giving an accurate reflection of that available from the plasma over the cell's maturation period. Once incorporated into the red cell, the folate remains stable and unaltered for the life of that particular red cell – usually 120 days. Thus the circulation containing a mixture of red cells of different ages reflects the plasma status over the previous four months.

In the prospective study of Kirke *et al.*,[17] 81 women had embryos affected by NTDs in the index pregnancy. Their plasma, and more importantly, their red cell folate (RCF) status was compared to 247 randomly selected controls. While the median folate was significantly lower in the former group, it was clear that only a small number, less than one fifth, had RCF levels that one would conventionally regard as being folate deficient or borderline folate deficient (*i.e.* <150 µg/L RCF). The same was true of the pattern with plasma folate. Most of these samples were taken during the first trimester or shortly thereafter. Thus one could apparently conclude that, over the previous four months, the plasma folate in these women was adequate to supply the bone marrow with enough folate to keep the red cell levels in the normal range. One would infer that the developing embryo was also not folate deficient at least in the conventional sense.

5.2 Maternal Malabsorption of Folate as a Cause of NTDs

Bower *et al.*[23] suggested that women who went on to have NTD affected pregnancies had a constituent inability to absorb dietary folate in the normal manner. The evidence for this was that women with a previous history of NTDs have low folate. A similar suggestion has been made by Neuhouser.[24] It would seem clear, if the aetiology of NTDs were through maternal intestinal malabsorption, that this would have to manifest itself in low plasma levels and subsequently low red-cell folate levels in NTD affected pregnancies. Clearly the study of Kirke *et al.*[17] would not support this contention, where they found that the vast majority of women who went on to have NTDs in that same pregnancy had plasma and red cell folate levels in the normal range.

5.3 Evidence That There Is Abnormal Folate Metabolism in NTD Affected Pregnancy

Kirke *et al.*[17] found normal folate status in most of 81 women who had an NTD affected pregnancy. Daly *et al.*[25] carried out further analysis on this same population. They found that while most of the affected pregnancies took place in the face of normal folate status there was a marked difference in risk depending upon red cell folate level. It was clear that women with levels at or close to the normal cut off for deficiency, *i.e.* <150 µg/L, had a very substantial chance of having a NTD affected pregnancy, of over 6 per 1000 births, while women with very good folate status of over 400 µg/L had a prevalence of less than 0.6 per 1000 births. The relationship in between was not linear with a very sharp decline in risk as red cell folate status increased. This would suggest that an event involved in producing an NTD affected birth has a greater or lesser chance of happening, depending upon folate status over the entire physiological range. Folic acid supplements of 400 µg per day (or foods fortified with these levels) would have the effect of moving all women ingesting these levels to greater than 400 µg/L. It would appear from the shape of the curve that there is a leveling off of risk at red cell folate levels of greater than 400 µg/L but numbers in the Daly *et al.*[25] study at this level were too low to be certain that there was no increased benefit. If there was, however, it would appear to be small in magnitude. It thus appears that supplements are optimising some folate responsive event by achieving folate status at the top end of the physiological range, rather than by treating deficiency.

5.4 What is the Aetiology of This Folate Response?

All one can say with certainty from the trials and case control studies is that there is a folate responsive event or events in pregnancy that sometimes gives rise to a NTD. The frequency with which this happens may be much greater than appears from existing prevalence studies because there is good evidence that many fetuses with NTDs are spontaneously aborted or lost in early pregnancy.[6] The Daly *et al.*[25] study would suggest that this folate responsiveness is over the normal physiological range. There are 58 folate dependent enzymes and processes in mammalian cells and several cellular transport systems. It is possible that one or more of these is critical in the closure of the neural tube. Folate dependent enzymes are essential for the *de novo* synthesis of purines and pyrimidines which go to make up DNA and RNA (Figure 2). Impairment of the biosynthesis of DNA, in particular, would be expected to result in reduced cell division. In fact in subjects who are either folate deficient or being treated with antifolate drugs such as methotrexate there is clear evidence of reduced cell division, this being most obvious in rapidly dividing cells such as red cells and seen clinically as anaemia.[26] The folate cofactors are also involved in providing so-called carbon one units for the methylation cycle. The methylation cycle provides methyl groups for a wide range of methyltransferases. These methyltransferases have widely varying functions, as diverse as the biosynthesis of lipids and hormones through to the methylation of DNA. The controls on the methylation cycle ensure that it works even during folate deficiency. However, severe folate deficiency does give rise to a reduction of the methylation cycle that can be seen clinically. A neuropathy occurs that is similar to the type seen when the cycle is interrupted by vitamin B_{12} deficiency, which is called clinically sub acute combined degradation (SCD).[27] This is because in both instances there is a reduction in the synthesis of one of the principal components of the myelin, a protein called myelin basic protein.[27] Myelin basic protein requires the involvement of a methyltransferase which in turn requires an adequate supply of methyl groups. When the methylation cycle is impaired in either vitamin B_{12} deficiency or severe folate deficiency, the myelin synthesised is unstable and breaks down causing the observed myelopathy and neuropathy which is known clinically as SCD. Similarly, inborn errors of metabolism that affect the methylation cycle, either directly or indirectly, cause SCD. Thus folate dependent enzymes are involved in two distinct processes: many varied methylation reactions on one hand and DNA biosynthesis and cell division on the other. Interruption of either process could be expected to have serious consequences and could well impair the proper closure of the neural plate causing either spina bifida or anencephaly.

6 Prevention of NTDS

The literature outlined above would indicate that there are probably a wide range of genetic variants of folate dependant enzymes. So far, only one has been identified as compromising folate metabolism. The variant is of the folate dependent enzyme 5,10-methylenetetrahydrofolate reductase (MTHFR), where there is a cytosine to thymine substitution at base 677 that reduces folate status by about 30%, irrespective of one's folate intake.[28] This variant is also a cause of NTDs;[29] however, it accounts for only 13% of NTDs. Since over 72% of NTDs can be prevented by folic acid, it seems that there are some 60% of genes that code for folate dependent enzymes or processes still unaccounted for. Notwithstanding whether or not they have been identified, it would seem that they respond to extra folic acid. Thus the question remains as to how this extra folic acid can effectively be delivered to women who may become pregnant. As discussed above the elective changes in food intake, even if this was to fortified food, are unlikely to be effective. This will also be true of supplements. It would thus appear that the only thing that will be effective is a general fortification of the diet with folic acid. This general fortification could either be mandatory or, if it was very extensive, say amongst all or most of the country's flour producers, it could be voluntary. In either event, fortification would produce a very wide range of intakes of folic acid. Even if one takes a staple such as flour which finds its way not only into bread but also into pasta, confectionery *etc.*, there are large differences in individual consumption, not just in women of child bearing age but in the general population who would of course be exposed to such increases in folic acid also. The range of this exposure depends upon the target figure that one is trying to achieve to protect against NTDs. If this was the amount currently recommended (400 µg/d), derived from the intervention trials, one would see much of the population having intakes of 600–800 µg/d with some getting well over 1000 µg/day. It is recognised that these higher levels of folic acid will treat the anaemia which results from vitamin B_{12} deficiency. The commonest cause of such deficiency is the autoimmune disease pernicious anaemia (PA). This condition is absent in younger people and would not be expected to occur in women of child bearing age.[26] However, its prevalence increases with age and, while it is low, it definitely occurs in more and more people as they get older. These cases of PA are usually picked up medically by the diagnosis of the anaemia, but this anaemia would not occur in patients on folic acid. This 'masking' of the anaemia is an issue because in preventing its diagnosis it allows the other main feature of PA, namely a neuropathy, to progress undiagnosed, presenting at a later stage where much of its affects are not corrected upon therapy with vitamin B_{12}.[2] This masking of PA is known to occur at relatively high intakes of folic acid. Levels of 400 µg used in supplements are believed not to do this. Exactly at what level of folic acid it will occur is difficult to say because the only data available are from studies conducted decades ago when folic acid was inappropriately used clinically to treat PA. Many feel that levels of 1000 µg/d will not cause masking.[30] However, others feel that it could happen at much lower levels.[31] In the absence of any clear-cut position, most people in this field feel that fortification, if it to be widespread or mandatory, should not expose many in the general population to levels above an extra 400 µg per day. The measure recently introduced in the US where the flour has been fortified with 140 µg/100 g is calculated to lead to a mean increase in the general population of 100 µg/d. While many feel that this target was too low to be effective there is now evidence that levels lower than the target of 400 µg/d will be effective for preventing many NTDs. Daly *et al.*[32] found that a folate status commensurate with the lowest rate of NTDs could be achieved certainly by an extra 200 µg/d and possibly by levels as low as 100 µg/d if taken habitually as they would be in fortified flour. Indeed a very recent study on the effect of the US measure has shown it to be very effective indeed.[33] Prior to fortification, a large cohort of the elderly in the US were found to have a prevalence of very low folate status (<3 µg/L) of nearly 20%. After fortification at what was considered by many to be a very inadequate amount (140 µg/100 g of flour) the prevalence of those with low folate status has fallen to 1.7%. It thus appears that very low levels of fortification may do a lot of good and represent no risk to the population as a whole, thus offering folic acid as a case study for fortification.

References

1　L. Wills, *BMJ*, 1931, **i**, 1059.
2　J.M. Scott and D.G. Weir, *Essays in Biochemistry*, 1994 Vol. 28, 63, Editor K.F. Tipton, Portland Press, London.
3　J. McPartlin, A. Halligan, J.M. Scott, M. Darling and D.G. Weir, *Lancet*, 1993, **342**, 148.
4　J.M. Scott and D.G. Weir, *Recent Advances in Obstetrics and Gynaecology*, 1998, **20**, 1.
5　N. Wald, *CIBA Found. Symp. 181*, 1994, Wiley, London.
6　J.M. Elwood, J. Little, and J.H. Elwood 'Epidemiology and control of neural tube defects', 1992, Oxford, Oxford University Press.
7　C. Bower, F.J. Stanley, M. Croft, N.H. De Klerk, R.E. Davis, and D.J. Nicol, *Br. J. Nutr., 1993*, **69**, 827.
8　C. Bower and F.J. Stanley, *Med. J. Aust.*, 1989, **150**, 613.
9　M.K. Heide, N.D. Bills, S.H. Hinrichs and A.J. Clifford, *J. Nutr.*, 1992, **122**, 888.
10　B.M. Hibbard, *Obstet. Gynacol. British Commonw.*, 1964, **71**, 529.
11　J.L. Fraser and H.J. Watt, *Am. J. Obstet. Gynecol.*, 1964, **89**, 532.
12　J.A. Prichard, D.E. Scott, P.J. Whalley and R.F. Haling, *J. Am. Med. Assoc.*, 1984, **211**, 1982.
13　R.W. Smithells, S. Sheppard, C. Schorah, M.J. Seller, N.C. Nevin and R. Harris, *Lancet*, 1980, **i**, 339.
14　R.W. Smithells, N.C. Nevin, M.J. Seller, S. Sheppard, R. Harris, A.P. Read, D.W. Fielding, S. Walker, S.J. Schorck and J. Wild, *Lancet*, 1983, **i**, 1027.
15　K.M. Laurence, M. James, M.H. Miller, G.B. Tennant and H. Campbell, *Brit. Med. J.*, 1981, **282**, 1509.
16　MRC Vitamin Study Group, *Lancet*, 1991, **338**, 131.
17　P.M. Kirke, A.M. Molloy, L.E. Daly, H. Burke, D.G. Weir and J.M. Scott, *Quart. J. Med.*, 1993, **86**, 703.
18　A.F Czeizal and J. Duddas, *New Engl. J. Med.*, 1992, **327**, 1832.
19　Department of Health, Folic acid and the prevention of neural tube defects, report from expert advisory group, Lancashire: Health Publications Unit, 1992.
20　CDC Recommendations for the use of folic acid to reduce the number of cases of spina bifida and other neural tube defects, *MMWR* 1991, **41**, 1.
21　A.M. Molloy, P. Kirke, I. Hillary, D.G. Weir and J.M. Scott, *Archives of Disease in Childhood*, 1985, **60**, 1.
22　J.L. Mills, J. Tuomilehto, K.F. Yu, N. Colman, W.S. Blaner, P. Koskela, W.E. Rundle, M. Forman, L. Toivanen and G.G Rhoads, *J. Pediatr.*, 1992, **120**, 863.
23　C. Bower, F. Stanley, M. Croft, N. de Klerk, R.E. Davis and D.J. Nicol. *Brit. J. Nutr.*, 1993, **69**, 827.
24　M.L. Neuhouser, S.A. Beresford, D.E. Hickok and E.R. Monsen. *Am. J. Med. Genet.*, 1998, **16**, 196.
25　L.E. Daly, P.M. Kirke, A. Molloy, D.G. Weir and J.M. Scott, *J. Am. Med. Assoc.*, 1995, **274**, 1698.
26　I. Chanarin, 'The Megaloblastic Anaemias', 2nd Edition, Blackwell Scientific Publications, Oxford.
27　J.M. Scott and D.G. Weir, *Lancet*, 1981, **ii**, 337.
28　A.M. Molloy, S. Daly, J.L. Mills, P.N. Kirke, A.S. Whitehead, D. Ramsbottom, M. Conley, D.G. Weir and J.M. Scott, *Lancet* 1997, **349**, 1591.

29   A.S. Whitehead, P. Gallagherm, J.L. Mills, P. Kirke, H. Burke, A.M. Molloy, D.G. Weir, D.C. Shields and J.M. Scott, *Quart. J. Med.*, 1995, **88**, 763.
30   Dickinson, *Quart. J. Med.* 1995, **88**, 357.
31   D.G. Savage and J. Lindenbaum, 'Balliere's Clinical Haematology', Editor S.M. Wickransinghe, Balliere Tindall, London, p.657
32   S. Daly, A.M. Molloy, J.L. Mills, M.R. Conley, L.J. Young, P.N. Kirke, D.G. Weir and J.M. Scott, *Lancet*, 1997, **350**, 1666.
33   P.F. Jaques, J. Selhub, A.G. Boston, P.W. Wilson and I.R. Rosenberg, *New Eng. Med.*, 1999, **19**, 1449.

Promoting Folic Acid to Women of Childbearing Age

L. Thorpe and M.M. Raats*

HEALTH EDUCATION AUTHORITY, TREVELYAN HOUSE,
30 GREAT PETER STREET, LONDON SW1P 2HW, UK
*(FORMERLY RESEARCH MANAGER, HEALTH EDUCATION
AUTHORITY), UNIT OF HEALTH CARE EPIDEMIOLOGY,
UNIVERSITY OF OXFORD, INSTITUTE OF HEALTH SCIENCES,
OLD ROAD, OXFORD OX3 7LF, UK

1 Introduction

Since July 1995 the Health Education Authority (HEA) has been funded by the Department of Health to educate the public and health professionals about the benefits of folic acid for preventing the development of neural tube defects (NTDs) in the unborn child. The HEA has not been promoting this important B vitamin for any other reason.

The campaign's work is based on sound scientific evidence, as detailed in the recommendations of a DH expert advisory group report,[1] and has consisted of a variety of approaches. These have included work with health professionals, education for young people, developmental and tracking research, educational materials, advertising and other media work for both women planning pregnancy and women of childbearing age.

An important element of the work for women of childbearing age was the development of the folic acid flash labelling scheme in 1997, which gave support and impetus to food retailers and manufacturers, both to highlight products already fortified with folic acid and to extend the range of such products. The scheme was developed in consultation with statutory bodies and industry representatives and has now been adopted by 18 food retailers and manufacturers, with the result that the two flashes, 'with extra folic acid' and 'contains folic acid', now appear on over 250 products fortified with folic acid.

The scheme is a good example of nutrition claims supporting an important public health initiative and helping to raise awareness of folic acid

175

advice. The scheme was not devised to stand alone, and is regarded as a successful element within a broad programme of education rather than as an end in itself.

2 Background

In July 1991, the Medical Research Council Vitamin Study was published,[2] the Chief Medical Officers of the UK then convened an expert advisory group to review the evidence relating to folic acid and its role in preventing neural tube defects. Their report, published in 1992,[1] recommended that women could reduce the chances of their baby being born with a neural tube defect by increasing their daily intake of folic acid in the following ways:

- eating more folate-rich foods
- eating more foods fortified with folic acid - especially breads and breakfast cereals
- taking a daily 400 µg supplement from the time they start trying to conceive until the 12th week of pregnancy.

Despite urging doctors and nurses to convey this information to the women in their care, studies continued to show that women's awareness of folic acid remained low, as did the number of those taking folic acid.[3,4] For this reason, the HEA was awarded £2.3m to run an integrated education campaign, which had the following specific objectives:

- to increase awareness of the importance of taking *additional* folic acid before conception and for 12 weeks into pregnancy, in the general female population and in influential professional groups;
- to increase the availability of fortified breads and breakfast cereals;
- to increase the number and availability of appropriate supplements, especially those which are licensed;
- to make fortified products and supplements more easily identifiable.

In all its communications, the HEA emphasised the need for women to take a daily 400 µg supplement if they were having unprotected sex and likely to become pregnant.

The campaign was informed by an advisory group comprising representatives of key professional, scientific, public health, policy, commercial and lobbying interests. The work of the campaign consisted of various complementary elements, including national and local professional education and support, public education, work with the commercial sector, research and evaluation, and partnerships. The overall aim of the work was to create a positive climate of change, within which women would be more aware of folic acid, more likely to see information about it, more able to buy it easily, and more likely to take it.

3 Public Education

The campaign's principal public audiences are women planning pregnancy, women of childbearing age who may have children in the future (whom the campaign came to term 'future planners') and young people. Activities used to reach them included advertising, media and public relations, publications and a freephone advice line.

Initially, the campaign focused its work on women planning pregnancy, the core target group that needed this new advice as quickly as possible. The key message for this group was to take a daily 400 µg supplement from the time they stopped using contraception until the end of the twelfth week of pregnancy.

The focus of the HEA's campaign broadened in its second year to include not only those women actively planning pregnancy, but also 'future planners' of pregnancy. This broadening of scope placed more emphasis on the dietary message to eat more breads and breakfast cereals fortified with folic acid.

3.1 Context

For the HEA's campaign to succeed, it was important to influence action in spheres other than public and professional education, in order to alter the context of women's lives in which the advice would be acted on. Throughout the campaign the HEA worked closely with the commercial sector, developing partnerships with food retailers and manufacturers, trade associations and regulatory bodies, the pharmaceutical industry and manufacturers of pregnancy testing and ovulation predictor kits.

There were three important considerations for the HEA's work with the food industry:

- fortification of foods with folic acid is voluntary and not mandatory in the UK;
- it is estimated that between 30 and 50% of pregnancies in England and Wales are unplanned – folic acid advice can only be fully acted on if women are actively planning pregnancy;
- attitudes to fortification of food vary in different European countries, and it can be difficult to make meaningful and direct health claims for folic acid on food packaging. HEA and other bodies have recommended that folic acid be made an exception to the health claims legislation in the UK.

The HEA's early discussions with food manufacturers and retailers at the beginning of the campaign had revealed a willingness to fortify further products with folic acid if the obstacles of low awareness of folic acid and low demand for folic acid-containing products could be overcome. There was interest in the HEA providing support with their work in this area, possibly by developing a folic acid symbol. However, it became clear early on that it would be very difficult to develop universal criteria for a single symbol that could be used on supplements, foods naturally rich in folic acid and fortified foods.

A number of factors then led the HEA to concentrate on the promotion of fortified foods – not only increasing their availability, but also making them more visible. Firstly, there was the need to encourage their consumption in women of childbearing age in order to increase levels of folic acid in women who experience unplanned pregnancies. Secondly, there was increasing evidence that synthetic folic acid is much more bioavailable than natural folate. Thirdly, supplements were already easily recognisable by virtue of their being only folic acid.

Very importantly, awareness of fortified breads and breakfast cereals as a source of folic acid was consistently low among both health professionals and the public, showing a clear need to increase awareness of foods fortified with folic acid. Women were also saying they wanted to be able to identify foods that had added folic acid[5] (see Table 1). The majority of health professionals are not experts in nutrition. Given that the fortified foods route is very important for reaching all women of childbearing age, it was a matter of concern that doctors were much more aware of vegetables as a source of folic acid than of breads and breakfast cereals, despite the fact that synthetic folic acid is more easily absorbed. Two-thirds of health professionals consistently named green leafy vegetables as rich in folic acid; in the last year of the HEA's health professional tracking research (1997), only 28% mentioned cereals and 19% mentioned bread.[5]

Table 1   *Food sources seen by health professionals to be rich in folic acid*

	1996	1997
Green/leafy vegetables	66%	65%
Vegetables	15%	29%
Cereals	18%	28%
Fruit	11%	19%
Bread	6%	19%
Liver	6%	6%

It was also clear that women themselves were in need of more education in this area (see Table 2). When asked to name foods fortified with folic acid, the findings were roughly similar to the 1997 health professional survey: 16% named bread and 30% named cereals. Encouragingly, knowledge was higher among recently pregnant women: 23% named bread and 46% named cereals. This group of women included the HEA's original core target group of women planning pregnancy, who were receiving more information about folic acid after the campaign's launch in February 1996. The 'don't know' category reduced between 1996 and 1997, but remained at one third of women generally, although only one-fifth of pregnant women.[5]

Table 2   *Foods perceived to be fortified with folic acid*

	1997 General Women's Survey (n=617)	1998 General Women's Survey (n=420)	1998 Pregnant Women's Survey (n=319)
Foods fortified with folic acid . . .			
Bread	**11%**	**16%**	**23%**
Breakfast cereals	**27%**	**30%**	**46%**
Milk	3%	4%	6%
Other	1%	3%	7%
None	13%	19%	15%
Don't know	45%	34%	22%

Question asked: 'Can you think of any foods which are fortified with folic acid, that is they have folic acid added to them by the manufacturer?'

3.2 Folic Acid 'Flash' Labelling Scheme

These factors encouraged the HEA to develop a food labelling scheme to help women identify foods fortified with folic acid, principally breads and breakfast cereals. The scheme has become known as the 'folic acid flash labelling scheme' (see Figure 1). The term 'flash' refers to a symbol featured on food packaging, often to highlight a particular ingredient.

'With extra folic acid' is a rich source claim under the UK's Food Labelling Regulations[6] and denotes that a product contains at least 50% of the RDA for a non-pregnant adult (i.e. at least 100 μg per serving). 'Contains folic acid'

Foods carrying this mark contain at least 50% of the daily amount of folic acid required by a non-pregnant adult in a daily serving (i.e. a minimum of 100 μg per daily serving)

Foods carrying this mark contain at least one-sixth of the daily amount of folic acid required by a non-pregnant adult in a daily serving (i.e. a minimum of 30 μg per daily serving)

Figure 1 *The folic acid flash labelling scheme*

denotes that a product contains at least one-sixth of the RDA for a non-pregnant adult (i.e. at least 30 μg per serving).

It was clear from the HEA's discussions with the food industry that a labelling scheme would be most likely to be adopted if it had support from industry leaders, if trade associations recommended it and if the logo was a single-colour design. Allowing the logo to be used flexibly, in terms of its size, colour and location, would also enhance the scheme's appeal. Crucially, the scheme would need to be underpinned by a programme of education.

Several visual concepts and types of wording were tested with focus groups of women, who preferred the disc shape, which they felt stood out better, and appeared more bold than a diamond or banner shape. Women widely preferred the term 'with extra' to 'added', and marginally preferred it to the term 'fortified'. They felt 'with extra' sounded more 'natural'.

The scheme was developed in consultation with the Department of Health, the Ministry of Agriculture, Fisheries and Food, the Local Authorities Co-ordinating Body on Food and Trading Standards (LACOTS), the Food and Drink Federation, the Association of Cereal Food Manufacturers, the Federation of Bakers, the British Retail Consortium, the Health Food Manufacturers Association and a range of individual manufacturers and retailers. LACOTS produced guidelines to assist their local local trading standards officers with ensuring legal compliance with the scheme.[7]

3.3.1 Criteria. Companies must sign a legal agreement with the HEA before they can use the scheme, although there is no licence fee for joining. Only products fortified with folic acid can carry the flash, and its use must accord with the UK's Food Labelling Regulations.

The HEA has offered additional literature and support to companies joining the scheme, which has proved popular with both food retailers and food manufacturers. Although the flash may be featured on other foods fortified with folic acid, the HEA's additional active support has been restricted to breads and breakfast cereals fortified with folic acid. Examples include featuring the flash on a press advertisement for future planners of pregnancy and developing a poster promoting bread as a source of folic acid.

4 Results

By May 1997, there were encouraging data showing increases in consumer awareness of folic acid,[5] and the scheme has proved popular with both food retailers and food manufacturers, who have responded very positively to the opportunity to promote their products in this way. By March 1999, 18 companies had signed up, representing over 250 products, and companies regularly include new products in the scheme.

Women's awareness of the scheme after one year is encouraging[5] (see table 3). Interestingly, pregnant women who recalled seeing the flash were the ones who were more likely to mention cereals as being fortified with folic acid when asked which foods are fortified. This would appear to indicate that the scheme is achieving its objective of raising awareness of these products as sources of folic acid.

Table 3 *Spontaneous (unprompted) and prompted awareness of folic acid flashes to identify foods fortified with folic acid*

	1997 General Women's Survey (n=617)	1998 General Women's Survey (n=473)	1998 Pregnant Women's Survey (n=319)
Spontaneous (unprompted) awareness of a flash	10%	14%	24%
Prompted awareness of either flash	15%	25%	39%
Prompted awareness of 'With extra folic acid' flash	7%	12%	22%
Prompted awareness of 'Contains folic acid' flash	7%	10%	12%

Questions asked: 'Have you seen or heard about a special symbol which displays information about folic acid on food packaging' and 'Can I check if you have seen either of these symbols at all?'

5 Conclusions

The uptake of this labelling scheme has been aided by the strong scientific evidence for the role of folic acid in preventing neural tube defects. It is possible that the other potential health benefits of folic acid have also influenced companies' decisions to join the scheme.

Overall, the scheme has helped to build recognition of folic acid and to maintain it in the public eye. It not only appears on food packaging but is also featured in retailers' literature. It has been a useful element of the comprehensive strategy to educate and increase awareness among the general public, and for reinforcing messages from health professionals, schools and supermarkets. However, it is not an end in itself. Most significantly, it does not describe why folic acid is being highlighted as a nutrient, and it is not possible to accompany the flash with meaningful and direct health claims about the role of folic acid in preventing neural tube defects, which would undoubtedly strengthen its capacity to educate.

In the HEA's 1998 survey of pregnant and recently pregnant women (see Table 4) 38% of women claimed to have taken folic acid preconceptionally, a 50–60% increase from the year before;[5] 65% of these women had planned their pregnancy. Many more women who claimed to have taken it in the first 12 weeks of pregnancy had not planned their pregnancies. The social class and age categories in the vitally important 'when trying for a baby' stage show that it is largely women aged over 25, and those of higher social class, who were more likely to take folic acid preconceptionally.

While the fortification of dietary staples with folic acid remains voluntary, this makes it especially important that the advice to eat more breads and breakfast cereals fortified with folic acid should become as much a part of well-woman advice as the advice to take a folic acid supplement if you are

Table 4 *Whether and when folic acid was reported to be taken by women in the Pregnant Women's Surveys*

	1997 Pregnant Women's Survey	1998 Pregnant Women's Survey
Took folic acid		
All respondents	73% (n=299)	76% (n=319)
Those that actively planned pregnancy	83% (n=166)	84% (n=165)
Unplanned pregnancy/left to chance	59% (n=133)	67% (n=154)
Social grade ABC1	82% (n=97)	85% (n=137)
Social grade C2DE	67% (n=196)	69% (n=181)
Aged under 25 years old	59% (n=97)	71% (n=80)
Aged over 25 years old	79% (n=202)	78% (n=238)
When trying for a baby		
All respondents	24% (n=299)	38% (n=319)
Those that actively planned pregnancy	39% (n=166)	65% (n=165)
Unplanned pregnancy/left to chance	5% (n=133)	9% (n=154)
Social grade ABC1	34% (n=97)	58% (n=137)
Social grade C2DE	19% (n=196)	24% (n=181)
Aged under 25 years old	14% (n=97)	14% (n=80)
Aged over 25 years old	28% (n=202)	47% (n=238)
In first 12 weeks of pregnancy		
All respondents	54% (n=299)	68% (n=319)
Those that actively planned pregnancy	57% (n=166)	76% (n=165)
Unplanned pregnancy/left to chance	50% (n=133)	59% (n=154)
Social grade ABC1	56% (n=97)	79% (n=137)
Social grade C2DE	52% (n=196)	60% (n=181)
Aged under 25 years old	51% (n=97)	64% (n=80)
Aged over 25 years old	55% (n=202)	69% (n=238)

Question asked: 'Thinking about your current/recent pregnancy, did you take folic acid or folates either before pregnancy or while you have been/were pregnant?' and 'In which of these situations, if any, have you taken more folic acid or folates? (when trying to become pregnant; when not trying to become pregnant but not taking precautions; when not trying to become pregnant and taking precautions; in the first 12 weeks of pregnancy; when trying to become pregnant and until the 12th week of pregnancy (only in 1997[81]); from the 13th week of pregnancy until after birth; after giving birth; none of these; don't know)

planning to become pregnant. However, there is a long way to go before doctors will be routinely giving folic acid advice as part of general health advice to women over the age of 16 (see Table 5). In the HEA's 1997 survey of health professionals,[5] only 2% claimed to do so, compared with the 71% who said they gave information about folic acid as part of health advice to women planning pregnancy. Ideally, all women should be told about folic acid and be eating more as part of their staple diet. This would offer some additional protection against risk of NTDs, should they conceive unexpectedly.

Table 5 *Health professionals' spontaneous (unprompted) mentions of type of general health advice given to women over 16 and should be given to women planning pregnancy and pregnant women (Base: all respondents except school nurses)*

	is given to women over 16*		should be given to women planning pregnancy		should be given to pregnant women	
	1996	1997	1996	1997	1996	1997
Alcohol	68%	52%	26%	58%	25%	61%
Diet generally/healthy diet	80%	84%	66%	76%	63%	77%
Exercise/fitness	32%	75%	14%	34%	13%	44%
Folic acid	0%	2%	55%	71%	36%	39%
Smoking	82%	77%	37%	75%	36%	77%

*Base: all who give advice to women/girls regarding a more healthy lifestyle generally
Question asked: 'Which general health advice do you give to ?' (spontaneous (unprompted); multiple answers possible)

References

1 Department of Health (1992) Folic acid and the prevention of neural tube defects. Report from an Expert Advisory Group. London: Department of Health.
2 The MRC Vitamin Study Group (1991) Prevention of neural tube defects: results of the Medical Research Council Vitamin Study. *The Lancet*, **238**,131.
3 Department of Health/Central Office of Information (1995) Folic Acid – Knowledge in the General Female Population. London: Department of Health. Unpublished.
4 Health Education Authority (1996) Awareness, Attitudes and Behaviour towards Folic Acid amongst women. London: Health Education Authority .
5 Changing Preconceptions. Volume 2. The HEA Folic Acid Campaign 1995–1998. Research Report. London: Health Education Authority (1998).
6 The Food Labelling Regulations 1996. Statutory Instruments 1996 No 1449. London: HMSO.
7 LAC 12 97 6. Folic Acid: Labelling and Claims Guidelines. London: Local Authorities Co-ordinating Body on Food and Trading Standards.

Bioavailability of $n-3$ Polyunsaturated Fatty Acids (PUFA) in Foods Enriched with Microencapsulated Fish Oil

J.M.W. Wallace[1], A.T. McCabe[1], M.K. Keogh[2], P.M. Kelly[2], W.S. Gilmore[1] and J.J. Strain[1]

[1] NORTHERN IRELAND CENTRE FOR DIET AND HEALTH (*NICHE*), UNIVERSITY OF ULSTER, COLERAINE, NORTHERN IRELAND BT52 1SA, UK
[2] TEAGASC, DAIRY PRODUCTS RESEARCH CENTRE, MOOREPARK, FERMOY, CO CORK, IRELAND

1 Introduction

Epidemiological studies have shown an inverse correlation between intakes of oily fish and the incidence of ischaemic heart disease.[1] Intakes of $n-3$ PUFA in the European Union (EU) are currently half that recommended by the EU Scientific Committee for Food.[2] Levels may be increased by incorporating microencapsulated fish oil into a wide variety of food products. The aim of the present study was to evaluate the bioavailability of microencapsulated $n-3$ PUFA incorporated into foods compared with an equal amount of fish oil. Twenty five healthy female volunteers (aged 20–26 years) were randomly assigned to one of two groups; one group received 0.9 g of $n-3$ PUFA/day as a fish oil capsule (capsule group), while the second group (food group) received an equal amount of $n-3$ PUFA/day from enriched foods (bread, biscuits and soup). Baseline and post supplementation platelet fatty acid composition were analysed using gas chromatography and plasma cholesterol (chol) and triglycerides (trigs) were assayed using colorimetric assays. Differences between the two groups were compared using Students t-test.

Change from Baseline

Fatty acid	Capsule Group ($n = 12$)	Food Group ($n = 13$)	P
20:4	-0.80 ± 1.12	-1.82 ± 0.82	0.263
22:5	0.78 ± 0.14	0.85 ± 0.17	0.445
22:6	0.75 ± 0.60	0.75 ± 0.30	0.356
Chol mmol/l	0.18 ± 0.17	0.23 ± 0.09	0.289
Trigs mmol/l	-0.04 ± 0.06	0.01 ± 0.06	0.917

Values are mean $\pm$ sem and are expressed as percentage of total fatty acids

There was no significant difference in the change in platelet eicosapentaenoic acid or docosahexaenoic acid, or in plasma cholesterol and triglycerides between the two groups following the intervention. The results from this study indicate that $n-3$ PUFA from microencapsulated fish oil enriched foods are as bioavailable as $n-3$ PUFA in a capsule. Fortification of foods with microencapsulated fish oil, therefore, offers an effective way of increasing $n-3$ PUFA intakes.

This research is funded by an EU shared cost project (FAIR-CT-95-0085).

References

1   D. Kromhout et al. *New England Journal of Medicine* 1985, **312**, 1205.
2   Scientific Committee for Food Nutrient and Energy Intakes of the European Community, Report of the Scientific Committee for Food, 31 series. Luxembourg 1993.

Effect of Nutrient Dense Foods and Physical Exercise on Dietary Intake and Body Composition in Frail Elderly People

N. de Jong[1], J.M.M. Chin A Paw[1], C.P.G.M. de Groot[1],
G.J. Hiddink[2] and W.A. van Staveren[1]

[1]DIVISION OF HUMAN NUTRITION AND EPIDEMIOLOGY,
WAGENINGEN AGRICULTURAL UNIVERSITY, DREIJENLAAN 1,
6703 HA WAGENINGEN, THE NETHERLANDS
[2]DUTCH DAIRY FOUNDATION ON NUTRITION AND HEALTH,
P.O. BOX 6017, 3600 HA MAARSSEN, THE NETHERLANDS

1 Introduction

The amount of research on physical frailty and its causes has begun to grow in recent years. In the literature, both broader and more narrow definitions of frailty have been used. One example is the definition of Buchner and Wagner,[2] in which physical frailty in elderly people is described as a state of reduced physiologic reserve associated with an increased susceptibility to disability. The downward process of frailty is often associated with inactivity, poor appetite, low food intake, sarcopenia and a reduction in physical functioning. This eventually results in an overall decline of the nutritional and health status. In order to investigate the effects of multiple micronutrient enriched foods and/or an all-round physical exercise program of moderate intensity on several indicators of this nutritional and health status in frail elderly people, a randomized placebo controlled trial was performed in The Netherlands in 1997.[3]

2 Methods

In total, 217 frail elderly people, who met the criteria of being 70 years of age or older, requiring home care or meals-on-wheels services, being inactive (no regular physical exercise) and having a BMI below average (≤ 25 kg/m^2), were randomized into one of the following intervention groups:

(a) nutrition: nutrient dense foods + social program
(b) exercise: 'regular' foods + exercise program
(c) combination: nutrient dense foods + exercise program
(d) control: 'regular' foods + social program.

The nutrient dense foods consisted of four types of dairy products and four types of fruit based products. Daily consumption of one dairy and one fruit product was obligatory and delivered daily ~100% of the Dutch RDA of the vitamins: D, E, B_1, B_2, B_6, folic acid, B_{12} and C and ~25–100% of the Dutch RDA of the minerals: calcium, magnesium, zinc, iron and iodine. The 'regular' products contained only the natural amount of micronutrients. Total energy content was the same for the regular and nutrient dense products: 0.48 MJ/day.

The all-round, moderately intense, progressive exercise program comprised exercises in a 45-minute group session carried out twice a week. The sessions were co-ordinated by skilled teachers. The social program, serving as a control for attention, comprised creative activities and lectures. Transport to and from all sessions was arranged. The intervention period was 17 weeks and data were collected at baseline and at the end of the study. Food intake was measured with a three-day (non-consecutive) dietary record, and body and bone composition with DEXA (dual energy X-ray absorptiometry).

3 Results

The study was successfully completed by 161 subjects. Valid body composition data were available for 143 participants. Mean age was 78±6 years and 70% of the study population were women. Mean BMI, at baseline, was 24.3±2.7 kg/m^2. Mean daily energy intake in men was 8.9±2.1 MJ, and in women 7.0±1.4 MJ. Investigating the changes after 17 weeks of intervention revealed no interaction between both interventions with respect to effects on intake and body composition. The non-exercisers (n = 68) showed a small decrease in energy intake (-0.3±1.8 MJ/day) after 17 weeks of intervention whereas the exercisers ($n = 75$) improved slightly (0.2±1.2 MJ/day). These changes were related with change in lean body mass. In Figure 1 a comparison of the differences (%) between the exercising group *versus* the non-exercising group and the enriched food group versus the regular food group are presented for bone density.

In conclusion, foods enriched with a physiological dose of multiple micronutrients slightly influence bone parameters in frail elderly people. Exercise seems important for the preservation of muscle mass and energy intake.

References

1 K. Rockwood, R.A. Fox, P. Stolee, D. Robertson and B.L. Beattie, *CMAJ*, 1994, **150**, 489.
2 D.M. Buchner and E.H. Wagner, *Clin. Geriatr. Med.*, 1992, **8**, 1.

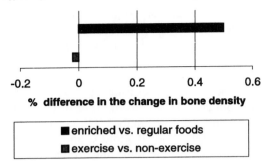

Figure 1 *Percentage difference in the change in bone density, over 17 weeks, between the enriched food group vs. the regular food group and the exercising group vs. the non-exercising group*

3   N. de Jong, J.M.M. Chin A Paw, C.P.G.M. de Groot, G.J. Hiddink and W.A. van Staveren, Dietary supplements and physical exercise affecting bone and body composition in frail elderly (Submitted).

Poster Presentations

Production of $n-3$ Enriched Eggs: New Promising Resources of Nutritionally Important Long Chain Fatty Acids

Heikki S. Aro[1], Tuomo Kiiskinen[2], Sini Panula[2], Tuomo Tupasela[1], Rainer Huopalahti[3], and Eeva-Liisa Ryhänen[1]

[1]AGRICULTURAL RESEARCH CENTRE OF FINLAND, FOOD RESEARCH, JOKIOINEN, FINLAND
[2]AGRICULTURAL RESEARCH CENTRE OF FINLAND, ANIMAL BREEDING, JOKIOINEN, FINLAND
[3]UNIVERSITY OF TURKU, DEPARTMENT OF BIOCHEMISTRY AND FOOD CHEMISTRY, TURKU, FINLAND

1 Introduction

Many studies have shown that the fatty acid content of eggs can be modified by changes in the diet of the hens. Therefore, it has been suggested that eggs can be an alternative to fish and oilseeds as a source of $n-3$ fatty acids.

Sensory analysis is often the only method to determine whether the product is acceptable. In many cases the sensory properties of eggs enriched with $n-3$ fatty acids have been recognised as undesirable.[1,2]

In this study we investigated the effects of dietary rapeseed oil, flax oil and fish oil on the fatty acid composition and sensory properties of eggs. Especially, we wanted to find any possibly organoleptic properties of modified eggs.

2 Materials and Methods

Altogether 72 White Leghorn hens were randomly divided to four groups. Hens were fed with test diets for 35 days and the eggs analysed were collected during days 30–35. Five yolks from each group were pooled, total lipids were extracted and the fatty acids were analysed as their methyl esters by gas chromatography.

In the sensory analysis, eggs were evaluated by two different methods. In the multiple comparison test, six panelists were asked to evaluate four different egg samples. One of the eggs was a reference sample. Panelists evaluated odour, taste and general acceptance of yolks using a scale: − 2 (clearly worse than the reference), − 1 (slightly worse than the reference), 0 (similar to the reference), +1 (slightly better than the reference), +2 (clearly better than the reference).

In a triangle test, panelists were informed that two out of three eggs were similar. Panelists were asked to compare the odour, taste and other sensory properties of the egg yolks and, according to these properties, find which egg was different. In every test, two of the eggs came from the control group and the third one came from a test group.

3 Results

1. In every dietary test group the fatty acid profile of yolk changed efficiently.
2. The amount of α-linolenic acid increased by a factor of 13 in the flax group, and in the fish oil group the amount of DHA by a factor of 2.5.
3. In the sensory evaluation the panelists considered that the taste of the rapeseed group of eggs was better than the control group (reference) eggs. The fish oil group got the lowest scores.

4 Conclusions

The fatty acid content of eggs can be easily modified by changes in dietary fat. Both fish oil and flax seed oil supplements in feed increase the proportion of the long chain $n − 3$ fatty acids in eggs. However, in particular, the fish oil supplement causes off-tastes in sensory evaluation. To eliminate these problems a combination of different oils should be investigated.

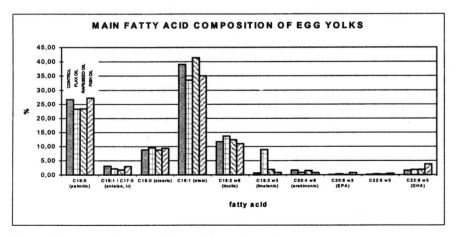

Figure 1. The fatty acid composition of egg yolk samples

TRIANGLE TEST		
	FLAXSEED	RAPESEED	FISH OIL
PANELIST 1	right	right	right
PANELIST 2	wrong	right	right
PANELIST 3	right	wrong	right
PANELIST 4	right	right	right
PANELIST 5	right	wrong	right
PANELIST 6	wrong	wrong	wrong

Figure 2 *The results of triangle tests*

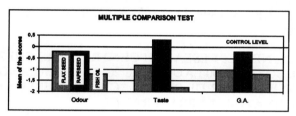

Figure 3 *The results of multiple comparison tests*

References

1 M.E. Van Elswyk, P.L. Dawson and A.R. Sams, *J. Food Sci.*, 1995, **60**(1), 85.
2 S. Leeson, L. Caston and T. MacLaurin *Poultry Sci.*, 1998, **77**, 1436.

Probiotic Effects Against Gastrointestinal Pathogens

A. Donnet-Hughes, S. Blum, D. Brassart, I. Corthésy-Theulaz,
A.L. Servin*, J.R. Neeser, H. Link-Amster, A. Pfeifer and
E.J. Schiffrin

NESTEC LTD., NESTLé RESEARCH CENTRE, VERS-CHEZ-LES-
BLANC, P.O. BOX 44, 1000 LAUSANNE 26, SWITZERLAND
*CJF 94.07 INSERM, UFR DE PHARMACIE, UNIVERSITé PARIS XI,
F-92296 CHâTENAY-MALABRY, FRANCE

The intestinal mucosal surface is at the interface between the host's immune system and the external environment. It is also the site through which pathogens try to invade the host and initiate infection. While the immune system operating at this mucosa provides strong defence mechanisms against pathogen attack, it is tolerant to a great number of commensal micro-organisms, which reside at the luminal surface. How the host discriminates between the commensal, symbiotic microflora and the exogenous pathogens is not fully understood though it is known that commensals are a strong ally to the host in the fight against infectious agents. Firstly, they provide a physical barrier that limits pathogen attachment to, and interaction with the mucosal surface. Secondly, they modulate host reactivity against the pathogen. Pro-biotic bacteria are a means of complementing the action of the microflora. Like commensal bacteria, they survive in the intestinal environment, attach to mucosal surfaces and modulate host immune functions.[1]

The major site of action of orally administered probiotic organisms remains a matter of conjecture, for changes in both small and large bowel have been reported in the literature using different probiotic strains.[2] By examining the effect of a single probiotic organism against pathogens that act at different levels of the gastrointestinal tract, we hoped to shed more light on this. This report describes the characteristics of one particular probiotic bacteria, *L. johnsonii* strain La1, to antagonise pathogens at different levels of the gastro-intestinal tract.

Pathological conditions of the stomach are associated with an overproduc-tion of stomach acids. However, treatments which counteract the increased gastric acidity lead to increased colonisation by the potentially pathogenic organism *Helicobacter pylori*. In fact, *H. pylori* is the most common bacterial infection in man and as such is a prime target for products containing probiotic organisms. Healthy volunteers with proven *H. pylori* infection were enrolled in a randomised, controlled, double-blind clinical trial.[3] They were treated for 2 weeks with a whey-based culture supernatant of *L. johnsonii* La1 combined with either omeprazole or with placebo tablets. Although the volunteers still showed signs of infection after treatment, breath tests indicated that the La1 cell culture supernatant had a partial, long-term inhibitory effect on *H. pylori* colonisation and that this inhibition was acid-independent.[3] Just as commensal organisms may provide a physical barrier to prevent pathogen attachment to intestinal tissues *in vivo*, previous work has shown that La1 can physically antagonise adherence of pathogens to gastrointestinal cells *in vitro*.[4] This study showed that at least with respect to *H. pylori* infection, extracellular products of La1 might contribute to the protection observed.

Pathogens, such as *Salmonella typhi*, invade the host by binding to specific receptors in the Peyer's patches of the small intestine. Colonisation of germ-free mice by La1 has been shown to improve survival rate in *Salmonella typhimurium*-infected mice.[5] However, simple blocking of specific receptors for Salmonella seems unlikely since the number of La1 given orally would be insufficient to block all the Salmonella binding sites. Nonbacteriocin, antibacterial products secreted by La1, may be involved.[5] However, it is also feasible that La1 stimulates host defence mechanisms. Volunteers ingesting milk fermented by La1, *Streptococcus thermophilus* and bifidobacteria, had increased serum IgA responses to an attenuated strain of *Salmonella typhi* Ty21a given in an oral vaccine [6] compared to controls who received only the vaccine.

Faecal colonisation is widely accepted as an indication of bacterial colonisation of the colon. Major components of the microflora in the large bowel include Enterobacteria, Clostridia and Bacteroides, which are considered as potentially pathogenic organisms. Indeed, *C. difficile* is a major cause of antibiotic-associated diarrhoea. Oral administration of La1 reduces faecal colonisation by *Clostridium perfringens*[7] without having any significant effect on the counts of Bacteroides or Enterobacteria. Given the established numbers and species of microorganisms present in the colon, probiotic organisms may be less effective against pathogens at this site. Direct interaction between the probiotic and host cells is certainly limited. Nevertheless, probiotic organisms do indeed influence the organisms present. Prokaryotic–probiotic interactions may predominate in the colon when an effective microflora is present, though soluble products may nevertheless modulate host cells. However, following antibiotic treatment, more direct interaction between host cells and probiotic organisms may afford additional protection against opportunistic pathogens.

To summarise, the probiotic bacteria *L. johnsonii* strain La1 can antagonise pathogens at multiple gastrointestinal sites. Precisely how this is achieved is unknown but extracellular bacterial products are likely to be implicated at all levels of the gastrointestinal tract by acting on both eukaryotic and prokaryotic cells. In the stomach and small intestine, which are not so highly colonised, direct contact between the probiotic and host cells may also contribute.

References

1   E.J. Schiffrin, D. Brassart, A.L. Servin, F. Rochat and A. Donnet-Hughes, *Am. J. Clin. Nutr.*, 1997, **66**, 515S.

2   R. Fuller, 'Probiotics the Scientific Basis', Chapman and Hall, London, 1992.

3   P. Michetti, G. Dorta, P.H. Wiesel, D. Brassart, E. Verdu, M. Herranz, C. Felley, N. Porta, M. Rouvet, A.L. Blum and I. Corthésy-Theulaz, *Digestion*, 1999, in press.

4   M.F. Bernet, D. Brassart, J.R. Neeser and A.L. Servin, *Gut*, 1994, **35**, 483.

5   M.F. Bernet-Camard, V. Liévin, D. Brassart, J.R. Neeser, A.L. Servin and S. Hudault, *Appl. Environ. Microbiol.*, 1997, **63**, 2747.

6   H. Link-Amster, F. Rochat, K.Y. Saudan, O. Mignon and J.M. Aeschlimann, *FEMS Immunol. Med. Microbiol.*, 1994, **10**, 55.

7   S. Blum, P.Serrant, F. Rochat, D. Brassart, A.M.A. Pfeifer and E.J. Schiffrin, (Abstract) International Symposium on Probiotics and Prebiotics, Kiel, Germany, June, 1998.

Hen Egg White Ovomucin, a Potential Ingredient for Functional Foods

J. Hiidenhovi,[1] M. Rinta-Koski,[2] A. Hietanen,[1]
S. Mantere-Alhonen,[2] R. Huopalahti[3] and E.-L. Ryhänen[1]

[1] AGRICULTURAL RESEARCH CENTRE OF FINLAND, FOOD
RESEARCH, 31600 JOKIOINEN, FINLAND
[2] UNIVERSITY OF HELSINKI, DEPARTMENT OF FOOD
TECHNOLOGY, 00014 UNIVERSITY OF HELSINKI, FINLAND
[3] UNIVERSITY OF TURKU, DEPARTMENT OF BIOCHEMISTRY
AND FOOD CHEMISTRY, 20014 UNIVERSITY OF TURKU, FINLAND

1 Introduction

One major research area in the functional food sector is to separate and characterise those food compounds that have bioactive properties, such as glycoproteins. One rich source of glycoproteins is hen egg albumen, which contains various glycoproteins like ovalbumin, ovotransferrin, ovomucoid and ovomucin. Ovomucin is a particularly interesting protein because it has been reported to have antiviral and antitumour properties.[1-3]

Ovomucin is easily fractionated from egg albumen by using isoelectric precipitation and centrifugation. However, the ovomucin precipitate thus formed is almost insoluble in water or conventional buffers. This limits its use as a food ingredient. In this study, some alternative methods, such as sonication and enzymatic hydrolysation, were studied to access whether they enhanced the solubility of ovomucin.

2 Materials and Methods

Ovomucin was prepared by means of the method developed by Kato et al.[4] with a few modifications as described in Hiidenhovi et al.[5] Solubility of ovomucin was studied using three different methods, namely stirring, sonication and enzymatic hydrolysis. After each solubility test, the amount of insoluble ovomucin was determined. Protein profiles of solubilized ovomucins were analysed by Superdex 200 HR gel filtration chromatography.

3 Results

1. Solubility of ovomucin was enhanced by the methods tested (Figure 1).
2. The composition of ovomucin varied depending on the method used to achieve solubility (Figure 2). The best solubility was achieved by enzymatic hydrolysis, which produced high molecular weight compounds (Figure 2).

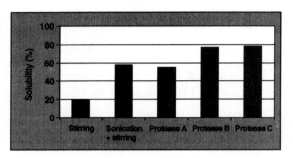

Figure 1 *The effectiveness of different dissolving methods for ovomucin*

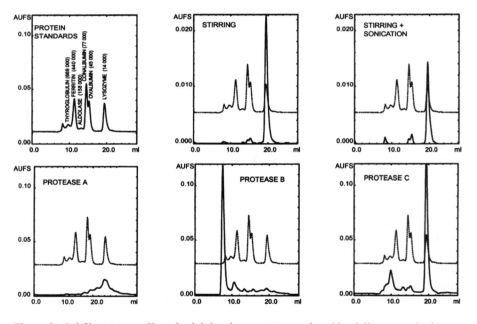

Figure 2 *Gel filtration profiles of solubilized ovomucins produced by different methods*

4 Conclusions

1. The results suggest that solubility of ovomucin can be improved by various methods.

2. Enzymatic hydrolysis seems to be an effective method for producing high molecular weight compounds from ovomucin.
3. In previous studies many beneficial effects of ovomucin on health has been reported. It can be suggested that soluble ovomucin is a potential ingredient for functional foods.

References

1   Y. Tsuge, M. Shimoyamada and K. Watanabe, *Biosci. Biotechnol. Biochem.*, 1996, **9**, 1503.
2   Y. Tsuge, M. Shimoyamada and K. Watanabe, *Biosci. Biotechnol. Biochem.*, 1996, **9**, 1505.
3   K. Watanabe, Y. Tsuge, M. Shimoyamada, N. Ogama and T. Ebina, *J. Agric. Food Chem.*, 1998, **46**, 3033.
9   A. Kato, R. Nakamura and Y. Sato, *Agric. Biol. Chem.*, 1970, **34**, 1009.
4   J. Hiidenhovi, H.S. Aro and V. Kankare, *J. Agric. Food Chem.*, 1999, **47**, 1004.

Wheat and Rye Brans: What Is the Difference?

S. Karppinen, K. Liukkonen, A.-M. Aura, P. Forssell and K. Poutanen

VTT BIOTECHNOLOGY AND FOOD RESEARCH, P.O.BOX 1500, FIN-02044 VTT, FINLAND

1 Introduction

Cereal brans are a good source of dietary fibre and may also offer some health effects *via* colon fermentation. Wheat bran has been suggested to reduce the risk of colon cancer and the slow fermentation rate has been suggested to be beneficial. The major polysaccharide of dietary fibre, both in wheat and rye bran, is arabinoxylan (pentosan). The aim of this study is to compare the fermentation patterns of wheat and rye brans.

2 Materials and Methods

Commercial rye and wheat brans obtained from Melia Ltd (Raisio, Finland) were milled to yield a particle size smaller than 1 mm. Starch and proteins were first digested enzymatically.[1] The digested brans (100 mg) were then fermented with human faecal inoculum *in vitro*.[2] The consumption of polysaccharides during the fermentation phase was followed by measuring the content of neutral sugars (arabinose, xylose, glucose) by HPLC after hydrolysis with sulphuric acid.[3] The progress of fermentation was followed also by measuring the formation of gases and short chain fatty acids (SCFA).

3 Results and Discussion

The total pentosan content of rye and wheat bran after digestion was about 35% whereas, in wheat bran residue the arabinose:xylose ratio was higher (Table 1). During fermentation, xylose was consumed more slowly than glucose (Figure 1). The fermentability of xylose is 64% in rye bran and 45% in wheat bran after 24 h fermentation. Slow fermentation also produces SCFA in the distal colon. The fermentability of glucose is 64% in rye bran and 52% in wheat bran. Arabinose was, in both cases, fermented only slightly (Figure 1). The total SCFA production was similar for both brans. Butyric acid, which is considered beneficial for the cells of the colon epithelium, was produced during both fermentations (Figure 2). The molar ratios of acetic, propionic and butyric acids after a 24 h fermentation period were 57:19:24 in rye bran and 57:22:21 in wheat bran. Rye bran fermentation produced slightly more gas (28 ml) than wheat bran (24 ml) (Figure 3).

4 Conclusion

The fermentation pattern of rye bran is rather similar to that of wheat bran, so similar health effects might be expected.

Table 1   *Content of the enzymatically digested brans*

	Rye bran, digested	Wheat bran, digested
Pentosan, %	36	34
Arabinose/xylose ratio	0,4	0,6
β-Glucan, %	5	3
Starch, %	4	3
Protein, %	14	12

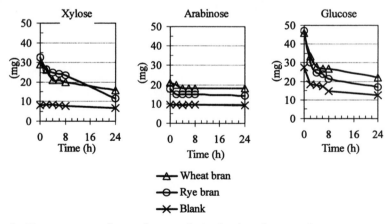

Figure 1 *Disappearance of neutral sugars during* in vitro *fermentation*

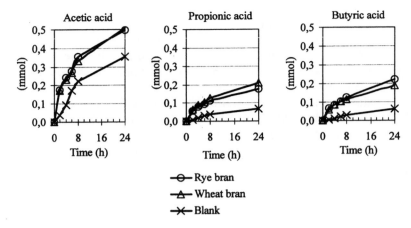

Figure 2 *Production of short chain fatty acids during* in vitro *fermentation*

References

1   A-M. Aura, H. Härkönen, M. Fabritius and K. Poutanen, *J. Cereal Sci.* (In press).
2   J-L. Barry, C. Hoebler, G.T. Macfarlane, S. Macfarlane, J.C. Mathers, K.A. Reed, P.B. Mortensen, I. Nordgaard, I.R. Rowland and C.J. Rumney, *Br. J. Nutr.*, 1995, **74**, 303.
3   V. Lebet, E. Arrigoni and R. Amado, *Lebensm. Unters. Forsch.*, 1997, **205**, 257.

Fermentation of Cucumber: Optimizing Process Technology for Design of Healthier Foods

T. Purtsi, M. Kilpi, B. Viander, H. Korhonen and E.-L. Ryhänen

FOOD RESEARCH, AGRICULTURAL RESEARCH CENTRE OF
FINLAND, FIN-31600 JOKIOINEN, FINLAND

1 Introduction

The interest in fermented vegetables has rapidly grown over the last few years. The positive effects of vegetables on health has been demonstrated in epidemiological studies. Investigations on the mechanism of bioactivity are still underway.[1-3] The aim of this study was to investigate and optimize the fermentation process of cucumbers. Commercial and isolated starter cultures were compared to spontaneous fermentation.

2 Materials and Methods

Cucumbers were fermented in 4.5% NaCl brine. The following cultures of lactic acid bacteria were used for fermentation: commercial *Lactobacillus pentosus*, *Lactobacillus plantarum*, *Pediococcus pentosaceus* and isolated lactic acid bacteria (strains A and B). Lactic acid bacteria, yeasts, moulds, *Enterobacteriaceae*, *Clostridium*-spores, pH and titratable acidity of brines were determined. Also, sensory evaluation, firmness of fermented cucumbers and microbiological quality of raw materials and end products were analysed. In the future, the probiotic properties of selected starter cultures will be determined *in vitro* and some selected products will be tested in clinical experiments.

3 Results

Starter fermentations were completed in about 7 days. In spontaneous fermentation, a duration of 14 days was determined (Figure 1). At the beginning of fermentation, the mixture containing isolated lactic acid bacteria decreased pH more effectively than the *Lactobacillus* starter (1 day and 3 days, respectively). However, the pH-level at the end of fermentation was lowest when the *Lactobacillus* starter was used. Differences in acid production were also determined (Figure 2). Best organoleptic quality and firmness of cucumbers was achieved in fermentation with the starter mixture containing isolated strains.

Hygienic quality of fermented cucumbers was good. No *Enterobacteriaceae* or *Clostridium* were found. According to the present study, good quality and freshness of one raw material are very important for a successful fermentation process. Yeasts may be a problem in cucumber fermentation. In this study, the count of yeasts in fermented cucumbers and brine varied between 10^3-10^6/ml, which is below the (official) limit value. No moulds were detected.

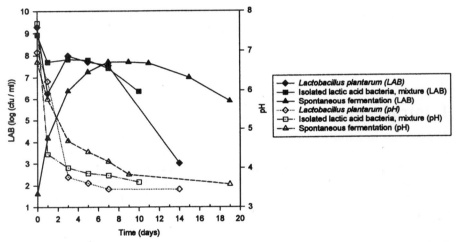

Figure 1 *Lactic acid bacteria (LAB) and pH of cucumber brine. (Fermentation was stopped when pH fell to about 4.0 and fermented cucumbers were transferred to +6°C)*

4 Conclusions

The results suggest that, by using starter cultures, fermented cucumber products can be manufactured more successfully and effectively than by a spontaneous fermentation process. Applying isolated strains of lactic acid bacteria as a mixture in the fermentation resulted in a product with good taste and structure. Fermented vegetables can be a good source of lactic acid bacteria. Some of these bacteria have been shown to have beneficial effects on health in humans and animals. The impact of isolated strains as probiotics and the possibility of beneficial effects on health will be investigated in further studies.

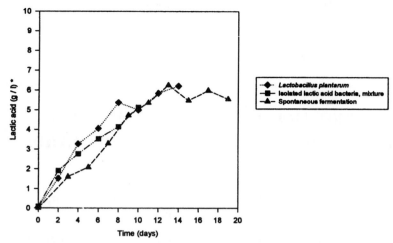

Figure 2 *Production of acid in cucumber brine (*determined by titration method and expressed as lactic acid, g/l)*

References

1   J.K. Collins, G. Thornton and G. O. Sullivan, *Int. Dairy J.*, 1998, **8**, 487.
2   G. Kalantzopoulos, *Anaerobe*, 1997, **3**, 185.
3   S. Scheinbach, *Biotechnol. Adv.*, 1998, **16**, 3, 581.

Glucosinolates and Their Breakdown Products in Sauerkraut and Sauerkraut Juice

Marianne Taipale, Britta Viander, Juha-Matti Pihlava, Hannu J. Korhonen and Eeva-Liisa Ryhänen

AGRICULTURAL RESEARCH CENTRE OF FINLAND, FOOD RESEARCH, 31600 JOKIOINEN, FINLAND

1 Introduction

Brassica vegetables have been shown to have several beneficial effects on human health. Some of these properties may be due to glucosinolates and their breakdown products that occur in cruciferous vegetables.[1,2] Even though there is currently no clear clinical evidence of the health benefits of fermented vegetables, they are perceived as functional foods by consumers.

2 Materials and Methods

In both fermentations the white cabbage (*Brassica oleracea* L. var. *capitata* L.) cv. *Lennox* (Bejo Zaden, The Netherlands) was used. Cabbage was cultivated at the Agricultural Research Centre of Finland (Piikkiö, 60 °23″ and 22 °33″) and harvested on 21.10.1998. Once harvested, cabbage heads were stored at +0.5 °C. Before fermentation, cabbage heads were stored at +2 °C for 24 h.

The first fermentation was performed with a lactic acid bacteria isolate at +20 °C. The second fermentation was spontaneous.

The first samples were collected from fermentation chambers for chemical analysis after mixing of salt and the starter-culture. The last samples were collected from fermentation chambers when the juice was pressed. Measurements of pH and microbiological analysis were performed on pressed juices.

Glucosinolates were determined according to standard procedures (BS 4325) with slight modifications. Briefly, the freeze-dried sample was extracted with 70% methanol at 85 °C for 10 min. The sample was centrifuged and a portion of the supernatant was applied to the DEAE Sephadex A25 ion exchange column with sulfatase. After overnight incubation, desulfoglucosinolates were eluted with water. Desulfoglucosinolates were determined using a gradient HPLC method. Chromatography was achieved using a Lichrosphere RP-18 (125 × 4 mm i.d.) column and an aqueous acetonitrile mobile phase at 30 ° C. Desulfoglucosinolates were monitored at 229 nm using a diode array detector. Injection volume was 200 µl.

Nitrate, nitrite and thiocyanate (SCN⁻) were determined using an isocratic HPLC method. Anions were based on an Asahipak ODP-50 column (125 × 4mm i.d.) and a buffer–acetonitrile (86:14) mobile phase. Identification and quantitation of the anions was based on indirect UV-absorption at 360 nm. The injection volume was 20 µl.

3 Results

Sensory evaluation indicated excellent organoleptic properties of fermented products. Furthermore, microbiological studies also indicated excellent properties.

From fresh cabbage, glucoiberin, progoitrin, sinigrin, glucoraphanin, gluconapin, 4-OH-glucobrassicin, glucobrassicin and 4-MeO-glucobrassicin were detetrmined. At the end of both fermentations, only 4-OH-glucobrassicin was detected. There was no significant difference in 4-OH-glucobrassicin concentration between fermentations.

Fermentation performed with a starter culture showed higher levels of nitrate than the spontaneous fermentation. For spontaneous fermentation, nitrate concentration in sauerkraut was 97 mg/kg F.W. and for the fermentation with starter culture it was 209 mg/kg F.W. The levels for sauerkraut juices were 109 and 207 mg/kg F.W. respectively. Nitrate levels in both fermentations were higher than in fresh cabbage (25 mg/kg F.W.). In all samples, the levels of nitrite and SCN⁻ were below the limit of detection (15 and 100 mg/kg F.W. respectively).

4 Conclusions

Most glucosinolates are decomposed during the fermentation process. Future experimentation will focus more on determination of breakdown products in fermented vegetable products.

During the fermentation of white cabbage to sauerkraut, nitrate levels were increased but the levels of nitrite remained low. These findings are in disagreement with data reporting nitrate concentrations in previous studies.[3,4] Discrepancies between current and published findings may be accounted for by differences in fermentation material or laboratory methods. Current findings suggest that nitrate content may be affected by starter bacteria. Based on the current data, both nitrite and nitrate concentrations do not represent a risk to human health.

References

1   D.T. Verhoeven, R.A., Goldbohm, G. Van Poppel, H. Verhagen and P.A. van den Brant, *Cancer Epidemiol. Biomarkers Prev.*, 1996, **5** (9), 733.
2   L.O. Dragsted, M. Strube and J.C. Larsen, *Pharmacol. Toxicol.*, 1993, **72** Suppl 1, 116.
3   T. Herod-Leszczynska and A. Miedzobrodzka, *Rocz. Panstw. Zakl. Hig.* (abstr.), 1992, **43** (3–4), 253.
4   M. Nabrzyski, R. Gajewska and G. Bossy, *Rocz. Panstw. Zakl. Hig.* (abstr.), 1989, **40** (3), 198.

Addition of Inulin to Breakfast Does Not Acutely Affect Calcium Metabolism

Ulla Teuri[1], Merja Kärkkäinen[2], Christel Lamberg-Allardt[2] and Riitta Korpela[1,3]

[1] VALIO LTD R&D CENTRE, PO BOX 30, FIN-00039 HELSINKI, FINLAND
[2] DEPARTMENT OF APPLIED CHEMISTRY AND MICROBIOLOGY, DIVISION OF NUTRITION, PO BOX 27, FIN-00014 UNIVERSITY OF HELSINKI, FINLAND
[3] FOUNDATION FOR NUTRITION RESEARCH, PO BOX 30, FIN-00039 HELSINKI, FINLAND

1 Introduction

Fructo-oligosaccharides (FOS) are food components that are already used in functional foods and which have been widely investigated. FOS consist of fructose units and a glucose unit, and they are not hydrolysed until they reach the colon. Longer chain FOS, which have up to 60 sugar units per molecule, are called inulins. FOS possess many health benefits. Enhanced absorption of calcium may be one of them and is at present under intensive research. The enhancement of calcium absorption has been demonstrated in some studies in humans,[1,2] although the effect has not been found in all studies.[3] The results of rat experiments have shown repeatedly that FOS increase calcium absorption at least in these animals.[4,5,6,7]

The parathyroid hormone (PTH) is an important regulator of calcium metabolism. If the serum calcium concentration falls, PTH is released from the parathyroid. PTH level increases calcium release from the bone and calcium absorption from the intestine and the kidneys. On the other hand, when serum calcium increases after ingestion of calcium, the serum PTH concentration decreases. It has been shown that the serum PTH level decreases when calcium is added to the diet, both chronically[8] and acutely.[9,10]

The aim of the present study was to investigate the acute effects on calcium metabolism of inulin ingested either with fresh calcium-enriched cheese or with a pure calcium salt.

2 Materials and Methods

2.1 Subjects

Fifteen healthy young women volunteered to participate in this study. The subjects were not taking any medication apart from oral contraceptives, which were used by all the participants. They gave their written informed consent to participate in the study, which was approved by the Ethical Committee of the Faculty of Agriculture and Forestry, University of Helsinki.

2.2 Methods

This study consisted of two successive parts, both of which had a two-period cross-over design. Before these two parts there was a control session with no cheese or calcium supplement at breakfast. During the first part, either 100 g cheese with inulin (15 g per portion) or 117 g cheese without inulin was ingested at breakfast, in randomized order. The inulin cheese contained 15 g inulin (Raftiline®, Orafti, Belgium) and 210 mg calcium per 100 g, and the control cheese contained no inulin and 180 mg calcium per 100 g. In the second part, a calcium supplement, with or without inulin (15 g), was similarily ingested. The supplement (Calcium-Sandoz, Sandoz, France) contained 210 mg calcium. Breakfast was light and included only 47 mg calcium, without cheese or a supplement. There was a one-week wash-out period between all the study sessions.

Before breakfast, between 8 a.m. and 9 a.m after an overnight fast, and 2, 4, 6 and 8 hours after breakfast, a blood sample was taken. The intact parathyroid hormone (iPTH), ionized calcium (iCa) and serum total calcium were measured. The area under the curve values ($AUC_{0-8/8}$) were calculated for iCa and serum total Ca concentrations and for iPTH, to express the average change during the 8 hour follow-up. AUCs were based on the changes from the baseline values. The absolute changes from the morning baseline values were also calculated at 6 and 8 hours.

Urine samples were collected for 24 hours before and 24 hours after breakfast. Calcium and creatinine in the urine were measured. The total amounts of calcium in the urine collected over 12 and 24 hours after breakfast were calculated. Lunch was consumed four hours after the baseline sample had been taken, and a snack eight hours after the baseline.

The ingestion of milk products on the day before each test day was controlled so that it was the same for all the sessions, and no other milk products were allowed. The diet on the test day was standardized and was the same in every study session. The diet consisted of basic food ingredients, and contained 260 mg calcium when cheese or supplement were not included.

Both parts were analysed separately as two-period cross-over studies. Repeated measures analysis of variance for cross-over designs was used to compare the effects of fresh cheese, with or without inulin, and the calcium supplement, with or without inulin.

3 Results

Inulin did not significantly increase the serum iCa compared to the corresponding study sessions without inulin. There was a tendency for the serum ionized calcium concentration to increase more when cheese containing inulin was ingested compared to the cheese without inulin ($p = 0.06$, Figure 1), but because a period effect was found, the first study period was analysed. During the first period there were no statistical differences between the groups receiving inulin-containing cheese or cheese without inulin. Inulin did not affect the serum total Ca concentrations.

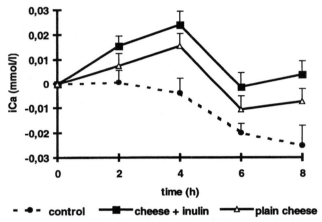

Figure 1 *Changes in serum ionized calcium concentration during the cheese sessions and the control session (mean ± SEM; n=15). No significant differences were found between the two cheese sessions*

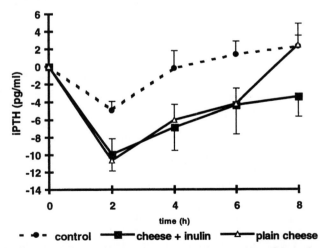

Figure 2 *Changes in intact serum parathyroid hormone concentration during the cheese sessions and the control session (mean ± SEM; n=15). No significant differences were found between the two cheese sessions*

Table 1   *Excretion of calcium (mmol; mean ± SD) during the study sessions at 12 h and 24 h (n=15). No significant differences were found between the cheese sessions or between the supplement sessions*

Excretion time	Control	Cheese + inulin	Cheese	Calcium supplement + inulin	Calcium supplement
12 h	0.77 ± 0.38	1.21 ± 0.44	0.93 ± 0.49	0.98 ± 0.42	0.99 ± 0.52
24 h	1.70 ± 1.99	1.60 ± 0.84	1.45 ± 0.90	1.47 ± 0.68	1.46 ± 0.72

The area under the iPTH concentration curve was not influenced by whether or not there was inulin with breakfast. However, eight hours after breakfast, the cheese containing inulin tended to keep the iPTH-level lower than the cheese without inulin (p = 0.08, Figure 2). Postload urinary calcium excretion was not significantly affected by inulin ingestion (Table 1).

4 Discussion and Conclusions

In this study, inulin ingestion did not acutely affect calcium metabolism. Earlier studies by isotope technique have shown the ability of inulin[1] and oligofructose[2] to increase calcium absorption in humans. Measuring serum PTH and ionized calcium does not provide direct information about calcium absorption, as do isotope techniques, but it does show the effects on acute calcium metabolism in the body, which is very important to bone health. It has been suggested that inulin may increase calcium absorption in the colon, rather than in the small intestine,[6,11] and therefore the time period allowed for our measurements may have been too short. The PTH concentration was lower with the inulin-containing cheese than with the control cheese 8 h after breakfast, and this may give a hint as to the later improvement in calcium balance. The study in ileostomic subjects, in which the calcium excreted from the small intestine was not altered by inulin ingestion, also supports the hypothesis of enhancement of calcium absorption in the colon.[12]

In the present study, no effect of inulin on urinary calcium excretion was noticed. Usually, the amount of urinary calcium excreted correlates well with the amount of calcium absorbed, as measured by the isotope technique.[13,14] The amount of calcium from cheeses at breakfast (210 mg) and of total calcium in the whole day's diet (470 mg) may have been too low to show any effect.

Fresh cheese enriched with 210 mg calcium and 15 g inulin per portion, ingested at breakfast, does not acutely affect the markers of calcium metabolism differently from cheese without inulin. Nor, according to the present study, did inulin appear to interfere with calcium metabolism.

References

1   C. Coudray, J. Bellanger, C. Castiglia-Delavaud, C. Rémésy, M. Vermorel and Y. Rayssignuier, *Eur. J. Clin. Nutr.*, 1997, **51**, 375.
2   E.G.H.M. van den Heuvel, T. Muys, W. van Dokkum and G. Schaafsma, *Am. J. Clin. Nutr.*, 1999, **69**, 544.
3   E.G.H.M. van den Heuvel, G. Schaasfma, T. Muys and W. van Dokkum, *Am. J. Clin. Nutr.*, 1998, **67**, 445.
4   R. Brommage, C. Binacua, S. Antille and A.L. Carrie, *J. Nutr.*, 1993, **123**, 2186.
5.   N.M. Delzenne, J. Aertssens, H. Verplaetse, M. Roccaro and M. Roberfroid, *Life Sci.*, 1995, **57**, 1579.
6   A. Ohta, M. Ohtsuki, S. Baba, T. Adachi, T. Sakata and E. Sakaguchi, *J. Nutr.*, 1995, **125**, 2417.
7   T. Morohashi, T. Sano, A. Ohta and S. Yamada, *J. Nutr.*, 1998, **128**, 1815.
8   G. Kochersberger, C. Bales, B. Lobaugh and K.W. Lyles, *J. Geront.*, 1990, **45**, M159.
9   J. Guillemant and S. Guillemant, *Calcif. Tissue Int.*, 1993, **53**, 304.
10   M.U.M. Kärkkäinen, J.W. Wiersma and C.J.E. Lamberg-Allardt, *Am. J. Clin. Nutr.*, 1997, **65**, 1726.
11   A. Ohta, Y. Motohashi, M. Ohtsuki, M. Hirayama, T. Adachi and K. Sakuma, *J. Nutr.*, 1998, **128**, 934.
12   L. Ellegård, H. Andersson and I. Bosaeus, *Eur. J. Clin. Nutr.*, 1997, **51**, 1.
13   J.E. Zerwekh, K. Sakhaee and C.Y.C. Pak, *Invest. Urol.*, 1981, **19**, 161.
14   W. van Dokkum, V. de la Guéronniére, G. Schaafsma, C. Bouley, J. Luten and C. Latge, *Br. J. Nutr.*, 1996, **75**, 893.

Fermentation of Cabbage and Carrot Juices by Using Isolated Lactic Acid Bacteria Strains: Development Towards Future Functional Foods

Britta Viander, Hannu J. Korhonen and Eeva-Liisa Ryhänen

AGRICULTURAL RESEARCH CENTRE OF FINLAND, FOOD RESEARCH, 31600 JOKIOINEN, FINLAND

1 Introduction

Fresh vegetables contain a varying bacterial flora, including spoilage bacteria and lactic acid bacteria. The original number of lactic acid bacteria in the raw material is normally very low and spontaneous lactic acid fermentation of vegetables is therefore a complex microbial process in which the lactic acid bacteria finally dominate the microbial flora.[1,2,3] By using lactic acid bacteria starters, the influence of the spoilage bacteria present in the raw material can be minimized. Furthermore, by using starters the fermentation time can be shortened.

2 Materials and Methods

Cabbage and carrot juices were fermented by using commercial starters and selected isolated lactic acid bacteria strains A and B. Different fermentation temperatures were used. Various cultivars of white cabbage and carrots were used and their suitability for fermentation processes was investigated.

During the fermentation process the following measurements were performed:

- pH
- microbiological analysis (lactic acid bacteria, enterococci, spores, yeast, moulds)
- chemical analysis (sugars, organic acids)

- organoleptical tests
- shelf-life.

3 Results

Using commercial starters and isolated lactic acid bacteria strains (mixture of A and B), cabbage juice (sauerkraut) has been successfully fermented. At the beginning of the fermentation, pH dropped very rapidly. The duration of starter fermentations was significantly shorter than spontaneous fermentations. The use of starters also minimized the negative influence of spoilage bacteria, resulting in improved hygienic quality.

Carrot juices have been fermented using a variety of isolated lactic acid bacteria strains and also using commercial starters. The impact of different fermentation temperatures on the fermentation process was evaluated. This study showed that the fermentation process can be significantly accelerated by using higher fermentation temperatures without adverse effects on organoleptical quality of products.

Various cultivars of both white cabbage and carrots were used and their suitability for fermentation processes was investigated. Cv. *Lennox* and *Lion* have been especially suitable for fermentation of cabbage juice and cv. *Fontana* has been very suitable for fermentation of carrot juice.

4 Conclusions

Cabbage and carrot juices can be fermented successfully by using commercial starters and, especially, by using selected isolated lactic acid bacteria strains. Juices with good microbiological, chemical and organoleptical quality were produced. Application of starter cultures can radically shorten fermentation processes which is of considerable economic importance. Potential beneficial impacts of the use of isolated lactic acid bacteria and fermented vegetable products on health will be investigated in future studies.

References

1   H.P. Fleming and R.F. McFeeters, *Biotechnol. Bioeng.*, 1988, **31**, 189.
2   R. Andersson, *Lebensm.-Wiss. u.-Technol.*, 1984, **17**, 282.
3   R.E. Andersson, C.E. Eriksson, B.A-C. Salomonsson and O. Theander, *Lebensm.- Wiss. u.- Technol.*, 1990, **23**, 34.

Quercetin and Rutin (Quercetin 3-*O*-Rhamnosylglucoside) Thermal Degradation in Aqueous Media under Alkaline Conditions

Dimitris P. Makris and John T. Rossiter*

DEPARTMENT OF BIOLOGICAL SCIENCES, WYE COLLEGE, UNIVERSITY OF LONDON, WYE, ASHFORD, KENT, TN25 5AH, UK

Abstract

Quercetin and its rutinoside, rutin, were chosen as model flavonols in order to investigate some basic principles of flavonol thermal degradation in alkaline aqueous media. Separation of the degradation products, as well as degradation kinetics under both oxidative and non-oxidative conditions, was accomplished by means of reversed-phase high-performance liquid chromatography (RP-HPLC), together with UV-Vis detection. Spectrophotometric monitoring of degradation in each case was also undertaken. The results obtained suggest rutin is much more resistant to degradation than its aglycone, quercetin. This stability was found to be dependent upon the occupation of the 3-*O* position by the sugar, although it is also partially attributed to the sugar moiety as such. Protocatechuic acid (3,4-dihydroxybenzoic acid) and, tentatively, phloroglucinol carboxylic acid (2,4,6-trihydroxybenzoic acid) were identified among the degradation products.

1 Introduction

Flavonols and their glycosides are very abundant, at considerable concentrations, in vegetables and fruits currently consumed by man.[1,2,3,4,5] Some plant-derived products, such as tea[6,7,8] and wine,[7] may also be rich flavonol sources.

In recent years flavonols have attracted much interest because there is sound evidence that they are absorbed by the mammalian organism in appreciable amounts,[9,10] and have been found to possess a wide range of biological activities, including effects on different oxidative metabolisms of arachidonic acid,[11] inhibition of cytochrome P-450 activity,[12] antithrombotic action,[13] blockage of human platelet aggregation and eicosanoid synthesis,[14] ability to inhibit procoagulant activity of adherent human monocytes,[15] inhibition of thyroid peroxidase,[16] and inhibition of xanthine oxidase.[17] In addition, flavonols along with other flavonoids have been characterised as "natural products with high pharmacological potency".[18]

In some cases, the levels of polyphenols decline considerably when plant foods and products undergo thermal processing. Common domestic processes such as boiling and frying can affect, to some extent, the flavonol content in onions,[5,19,20] tomatoes, lettuce, celery,[5] and broccoli florets.[21] Also, cooking can cause a decrease in total polyphenols of chickpea and blackgram,[22] tannins in pulses[23] and legume seeds[24,] and isoflavones in soybean[25] and soy products.[26] An important decline during drying has also been reported for plum anthocyanins and flavonol glycosides,[27,28] grape pomace polyphenols,[29] and oak leave tannins.[30] However, studies concerning changes in flavonoids, particularly flavonols, during processes involving heating, are scarce and elements regarding a detailed explanation of flavonoid loss are omitted, although there is enough evidence to support the idea that flavonoids indeed undergo degradation. Little is known about flavonoid thermal breakdown, and the mechanisms, as well as the structure, of possible degradation and reaction products remain to be elucidated. To further clarify the nutritional value and the beneficial effects of flavonols, as well as their role in food browning and deterioration, basic knowledge is required on the precise effect that thermal processes may have on these compounds.

Nevertheless, plant foods and products are very complex systems with respect to their chemical composition, and the study of flavonol thermal degradation in such systems would be, if not impossible, very difficult. For this reason, it was judged necessary that the examination of flavonol thermal degradation should be carried out on the basis of model systems. The alkaline pH used in this study represents a first approach. Quercetin and its rutinoside, rutin, were chosen because they are ubiquitous flavonols in edible plants, and possess many well studied biological activities. The rates of degradation and degradation products formation were determined by means of a new simple RP-HPLC method and UV-Vis spectroscopy, with very satisfactory specificity, repeatability, linearity, and sensitivity. Where possible, degradation products were also identified. Studies under low temperature were also undertaken, in order to better illustrate the role of heating and its impact on flavonol degradation.

2 Materials and Methods

2.1 Chemicals

Water used for HPLC analyses was distilled, purified by *EasyPure*® ultrapure water system, and filtered through 0.45 μm filters (*Millipore*). Acetonitrile, chloroform (CHCl₃), dichloromethane (CH₂Cl₂), diethyl ether, ethyl acetate, formic acid, hexane, tetrahydrofuran (THF) and methanol were from *BDH Laboratory Supplies* (Poole, England). Acetonitrile was of HPLC grade. All other solvents were of analytical grade. Catechol, lithium-aluminium hydride (1 M in THF), 3,4-dihydroxyphenylpropionic acid (dihydrocaffeic acid) and 2,4,6-trihydroxybenzoic acid (phloroglucinol carboxylic acid) were from *Aldrich Chemical Company, Inc.* (Milwaukee, USA). Quercetin, 3,4-dihydroxybenzoic acid (protocatechuic acid), and rutin (quercetin D-glucose, anhydrous magnesium sulfate ($MgSO_4$), L-rhamnose and phloroglucinol were from *BDH Laboratory Supplies* (Poole, England). 3,4-dihydroxyphenylpropanol was not commercially available, and thus synthesised as described below.

2.1.1 Synthesis of 3,4-dihydroxyphenylpropionic acid methyl ester. 1.98 g of 3,4-dihydroxyphenylpropionic acid (dihydrocaffeic acid) (10.8 mmol) was dissolved in 50 mL of methanol and 5 drops of concentrated sulfuric acid were added. The mixture was refluxed for 3 h while being monitored by TLC on silica plates (*Merck*), with ethyl acetate:hexane (1:1) as eluent. After the completion of the reaction (one single spot on TLC plate under UV light), the solvent was reduced *in vacuo* to approximately 5 mL, and made to 10 mL with ethyl acetate. The ethyl acetate solution was dried over $MgSO_4$, filtered through *Whatman* No. 1 paper filters, evaporated to dryness *in vacuo*, dissolved in 25 mL CH₂Cl₂, and mixed with activated charcoal. The charcoal-treated solution was filtered through *Whatman* No. 1 paper filters, concentrated and dried over P_2O_5 under high vacuum to give a white powder. The purity of the product was confirmed by TLC and RP-HPLC analysis, and its structure by ¹H NMR: 2.577 (2 H, *t*, -CH₂C=O, J = 7.00 Hz), 3.679 (3 H, *s*, -OCH₃), 2.861 (2 H, *t*, Ph-CH₂-, J = 7.00 Hz), 6.754 (H, *d*, J = 8.10 Hz), 6.591 (H, *dd*, J = 8.10 Hz), 6.739 (H, *d*, J = 2.06).

2.1.2 Synthesis of 3,4-dihydroxyphenylpropanol. Approximately 400 mg (2.04 mmol) of 3,4-dihydroxyphenylpropionic acid methyl ester were dissolved in 10 mL of anhydrous THF and transferred portionwise during 15 min, under nitrogen atmosphere, into an ice-cooled 250 mL, triple-neck, round-bottom flask containing 10 mL of LiAlH₄ (1 M in THF) and 10 mL of anhydrous THF. The solution was refluxed for 6.5 h, ice-cooled, and the excess of hydride destroyed by careful addition of 20 mL of water and 20 mL of 10% HCl. The organic phase was removed *in vacuo*, and the aqueous phase was extracted three times with 50 mL of diethyl ether. The extracts were combined and dried over $MgSO_4$, and the solvent was evaporated *in vacuo* to yield a brown oily residue. This residue was applied on a silica gel 60 (*Merck*) column, and eluted successively with ethyl acetate:hexane (1:1), ethyl acetate:hexane (3:1) and CHCl₃:methanol (8:2), to give 15 fractions of approximately 15 mL. Fractions were chromatographed on a silica TLC plate (*Merck*) with ethyl acetate:hexane (1:1) as eluent, and samples giving the same spots with the same R_f under UV-light (254 nm), were combined and concentrated, to yield a yellowish oil. The purity of the product was confirmed by TLC and RP-HPLC analysis, and its structure by ^{1}H NMR and mass spectrometry: *m/z* 168 (50%), 150 (6), 136 (10), 123 (100), 91 (35.5), 77 (46), 43 (72), 27 (44); 8.622 (3,4 -OH, *s*), 1.664 (2 H, Ph-CH$_2$-CH$_2$-, *m*), 2.473 (2 H, Ph-CH$_2$-, *t*, J = 1.65), 3.432 (2 H, CH$_2$-OH, *dd*, J = 6.04, 6.31), 6.441 (H, *dd*, J = 1.92, 7.97), 6.650 (H, *d*, J = 7.97), 6.603 (H, *d*, J = 1.92).

2.2 Treatment of Flavonol Solutions Under Non-Oxidative Conditions

2.2.1 High temperature (97 °C). A 100 mL, triple-neck, round-bottom flask, equipped with a water-cooled reflux condenser, was used. Heating was made by means of a magnetically stirred oil bath.

50 mL of a 0.2 M KOH solution (1.12%), pH 13.0 ± 0.1, containing 1.65 mM quercetin were heated at 97 °C. The KOH solution was degassed with argon prior to analysis, in order to eliminate oxygen. Samples were taken after 30, 60, 120 and 240 min of boiling. Treatments were performed in triplicate. Data reported are mean values. Solutions containing rutin (quercetin 3-*O*-rhamnosylglucoside), and quercetin/L-rhamnose/D-glucose (millimolar ratio 1:1:1) were treated similarly.

2.2.2 Room temperature (20 ± 1 °C). Treatments were carried out under argon stream introduced into the flask through a paraffin-containing trap, at room temperature (20 ± 1 °C). Mixing of the solution was accomplished by means of magnetic stirring.

2.3 Treatment of Flavonol Solutions Under Oxidative Conditions

2.3.1 High temperature (97 °C). The flask used was that described in paragraph 2.2.1. Heating was provided with a magnetically stirred oil bath. Oxidative conditions were established by bubbling air into the solution during boiling, through a glass tube connected with a pump. The rate of bubbling was constant throughout treatment, and as low as to maintain boiling.

2.3.2 Room temperature (20 ± 1 °C). Mixing of the solution was done by means of magnetic stirring. Oxidative conditions were established as described previously.

2.4 Reversed-Phase High-Performance Liquid Chromatography (RP-HPLC) Analysis

2.4.1 Sample preparation. 10 mL of sample was withdrawn from the reaction flask and placed in a 10 mL volumetric flask. The sample was immediately cooled with tap water (in the case of high-temperature treatment) and adjusted carefully to pH 3.0 with concentrated HCl. The pH-adjusted sample was concentrated in a rotary vacuum evaporator ($T \leq 40\,°C$), made to 10 mL with 50% aqueous methanol (in cases where insoluble material was observed, 70% aqueous methanol was used), and filtered through *Millex-HV$_{13}$*, 0.45 µm, syringe filters (*Millipore*). This solution was used for HPLC analysis. Precautions were taken to avoid contact with air throughout preparation. Analyses were carried out immediately after preparation.

2.4.2 Analytical HPLC procedure. A *Waters 600E* and a *Phillips PU 4100* liquid chromatographs were used. Detection was made by means of an *Applied Biosystems 757* detector set at 260 nm, and a *Philips PU 4021* diode array detector. Both systems were computer-controlled. The column was a *Waters Symmetry* C$_{18}$, 3.9×150 mm, 5 µm, protected by an *OptiGuard* guard column C$_{18}$, 3 mm. Columns were thermostatically controlled to maintain a temperature of 40 °C. Eluents were (A) water (pH 2.5 with orthophosphoric acid) and (B) acetonitrile:eluent A (6:4), and the flow rate was 1 mL/min. Injection was made by means of a Rheodyne injection valve with 20 µL fixed loop. The elution programme used is shown in Table 1. The column was washed with 100% acetonitrile and re-equilibrated with 100% eluent A before the next injection.

Identification of quercetin and rutin was based on retention times of the original compounds and spectral data. Quantitation was made by external standard method. Standard solutions were prepared in methanol or absolute ethanol, and kept at $-20\,°C$. Evaluation of the method is illustrated in Table 2.

Table 1   *RP-HPLC elution programme used for quercetin and rutin determination*

Time (min)	Eluent A (%)	Eluent B (%)
0	100	0
3	100	0
63	60	40
73	60	40

Table 2   *Evaluation of the RP-HPLC method used for quercetin and rutin analysis*

Compound	LOQ (µg/mL)[b]	Day-to-day variation (%)[b]	Linearity (r^2)[c]
Quercetin	0.142	1.49	0.9984
Rutin	0.174	1.51	0.9976

[a] LOQ: Limit of quantitation at 260 nm.
[b] Linearity: Given as correlation coefficient (r^2) over a concentration range varying from 0.094 to 1.497 mM for quercetin and from 0.117 to 1.871 mM for rutin.
[c] Day-to-day variation: Given as %RSD (relative standard deviation).

2.5 UV-Vis Spectroscopy

2.5.1 Sample preparation. This was as described for the analytical HPLC procedure. Samples were diluted 1:10 with methanol prior to analysis, unless elsewhere specified. All samples were analysed shortly after preparation.

2.5.2 Analytical procedure. A *Shimadzu* UV-Vis scanning spectrophotometer, interfaced with a UV-2101PC software was used. Absorbance was read at 420 nm, and spectra were obtained throughout a wavelength range varied from 230 to 600 nm, in a 1.0 cm path length cuvette.

2.6 ^{1}H NMR Spectroscopy

The ^{1}H NMR spectrum of protocatechuic acid was obtained by a *Bruker WM 250* spectrometer. CD_3OD was used as solvent and TMS as internal standard. The spectra of 3,4-dihydroxyphenylpropionic acid methyl ester and 3,4-dihydroxyphenylpropanol were obtained with a *JEOL GSX270 FT* NMR spectrometer with auto sample changer attachment, with $CDCl_3$ and DMSO as solvents and the same internal standard.

2.7 Mass Spectrometry

Low resolution spectrum of protocatechuic acid was obtained by a *Kratos MS890* (Kratos Ltd., Manchester, UK), under electron ionisation conditions at a source temperature of 200 °C. Samples were inserted onto the direct insertion probe which was then heated from 40–500 °C at 50 °C/min, while acquiring spectra at 10 sec/decade. For 3, 4-dihydroxyphenylpropanol, a *VG ZAB-2F* double focusing mass spectrometer was employed.

2.8 Large Scale Treatments

2.8.1 Procedure A. 400 mL of degassed 0.2 M KOH, pH 13.0 ± 0.1, containing 1.65 mM quercetin were introduced into a 500 mL double-neck, round-bottom flask, equipped with a water-cooled reflux condenser and a stopper, and the solution was boiled for 240 min. After the completion of the treatment, concentrated HCl was added until a colour change was observed, and the sample was allowed to cool down to ambient temperature. Subsequently, the sample was adjusted to pH 3.0 with concentrated HCl, evaporated to dryness in a rotary vacuum evaporator, dissolved in approximately 30 mL methanol, and filtered through a *Whatman* No 1 double filter paper. The filters were washed with methanol, and the filtrate was evaporated to dryness as described previously, re-dissolved in 5 mL methanol acidified with formic acid, and filtered through *Millex-HV$_{13}$* 0.45 μm syringe filters (*Millipore*). This solution was used for semi-preparative RP-HPLC. Rutin solutions were treated similarly.

2.8.2 Procedure B. 400 mL of degassed 0.2 M KOH, pH 13.0 ± 0.1, containing 1.65 mM quercetin were introduced into a 500 mL, double-neck, round-bottom flask equipped with a stopper and a paraffin-containing trap. The solution was stirred for 400 min at 20 ± 1 °C, under a continuous argon stream. The following steps were as described previously.

2.8.3 Procedure C. 400 mL of 0.2 M KOH, pH 13.0 ± 0.1, containing 1.65 mM quercetin were introduced into a 500 mL double-neck, round-bottom flask equipped with a paraffin-containing trap and a glass tube connected with a pump. The solution was stirred for 300 min at room temperature, under a continuous intense bubbling of air. The following steps were as described previously.

2.8.4 Semi-preparative RP-HPLC procedure. The column used was a *Waters Spherisorb®* S5 ODS2 C_{18}, 10 mm × 250 mm, 5 μm, protected by an *OptiGuard* guard column, C_{18}, 3 mm, and maintained to a temperature of 40 °C. Eluents used were (A) water (pH 2.5 with formic acid) and (B) acetonitrile:eluent A (6:4) and the flow rate was 2.5 mL/min. Injection was made by means of a Rheodyne injection valve with a 500 μL fixed loop. The elution programme was the one used in the analytical HPLC procedure, although other gradients with minor modifications were also used.

The following repeated injections, peaks were collected manually, taken to dryness in a rotary vacuum evaporator, and re-dissolved in 1.5 mL MeOH acidified with formic acid. The solvent was evaporated with a stream of argon, and the residues were dried over P_2O_5 under high vacuum, and stored at − 20 °C until analysed. Peak purity was checked by using the analytical HPLC procedure. Impure peaks were re-purified by the same semi-preparative HPLC method.

3 Results and Discussion

3.1 Examination of Quercetin Solutions

3.1.1 Non-oxidative conditions. As shown in Figure 1, a decrease of almost 83% in the initial concentration of quercetin occurred after only 30 min of boiling. However, complete degradation required more than 120 min. The rate of quercetin disappearance at 20 ± 1 °C was slower, but treatment resulted in a decomposition of approximately 95%.

The HPLC analysis showed that quercetin breakdown resulted in the formation of five compounds at high temperature and five compounds at low temperature, as illustrated in Figures 2A and 3A, and Table 3. Quercetin degradation was monitored by means of UV-Vis spectroscopy, and showed a progressive increase in absorbance at 287 and 297 nm for the high and the low-temperature treatments, respectively (Figures 4A and 4B), and did not

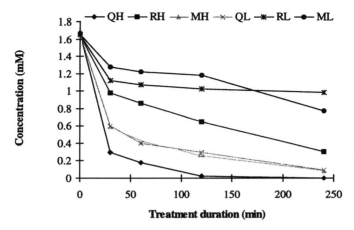

Figure 1 *Evolution of quercetin (Q), rutin (R), and quercetin/L-rhamnose/D-glucose (M) concentration during treatment under non-oxidative conditions. Letters H and L indicate treatment at high and low temperature, respectively*

Table 3 *% Relative peak area (number in parenthesis), recorded at 260 nm, of the main peaks detected during non-oxidative and oxidative degradation of quercetin and rutin, after 240 and 80 min, respectively*

Non-oxidative conditions				Oxidative conditions			
Quercetin		Rutin		Quercetin		Rutin	
HT	LT	HT	LT	HT	LT	HT	LT
QHR 1 (21.7)	QLR 1 (3.1)	RHR 1 (19.2)		QHO 1 (42.4)	QLO 1 (4.1)	RHO 1 (45.5)	RLO 1 (12.0)
QHR 2 (4.2)	QLR 2 (3.4)	RHR 2 (52.9)		QHO 2 (0.0)	QLO 2 (7.4)	RHO 2 (32.2)	RLO 2 (11.1)
QHR 3 (30.4)	QLR 3 (24.6)	RHR 3 (12.5)		QHO 3 (15.4)	QLO 3 (1.5)	RHO 3 (22.3)	RLO 3 (23.3)
QHR 4 (37.3)	QLR 4 (64.6)	RHR 4 (8.4)		QHO 4 (42.3)	QLO 4 (51.1)		RLO 4 (33.2)
QHR 5 (6.5)	QLR 5 (4.3)	RHR 5 (7.1)			QLO 5 (9.0)		RLO 5 (20.4)
					QLO 6 (27.1)		

reveal the existence of any other compound(s) absorbing beyond 300 nm. Among the compounds detected, protocatechuic acid and phloroglucinol carboxylic acid could easily be identified by comparison of their respective UV spectra and HPLC retention times with those of known standards. For protocatechuic acid, ^{1}H NMR and mass spectrometric studies were also undertaken (Table 4).

Table 4   *Mass spectrometric and ¹HNMR data of protocatechuic acid*

EI-MS data (m/z)	¹H NMR data (ppm)
154, 137, 109, 44	7.366 (1-H)
	7.360 (5-H)
	6.827 (4-H)

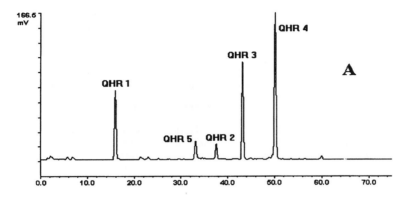

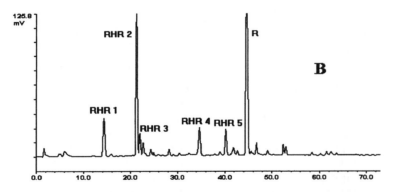

Figure 2 *RP-HPLC profile of quercetin (A), and rutin (B) degradation products after 240 min of boiling (97°C). Analysis of a 20 μL sample. Column: Waters Symmetry, C₁₈; mobile phase: (a) water (pH 2.5 with orthophosphoric acid), (b) acetonitrile: (a) (6:4). Detection at 260 nm. A gradient from 100% a to 60% a in 60 min was applied. QHR 1 (RHR 1): 3, 4-Dihydroxybenzoic acid (protocatechuic acid); QHR 2 (RHR 4), QHR 3 (RHR 5), QHR 4, QHR 5, RHR 2, and RHR 3: Unidentified degradation products*

Phloroglucinol carboxylic acid and QLR 3 were not formed during boiling (Table 3). On the other hand, QHR 5 and QHR 4 were not detected during treatment at 20 ± 1 °C, whereas QLR 4 (QHR 2) was by far the most abundant compound. Additionally, protocatechuic acid formation at 20 ± 1°C was

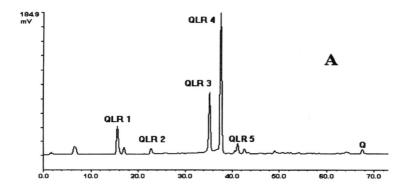

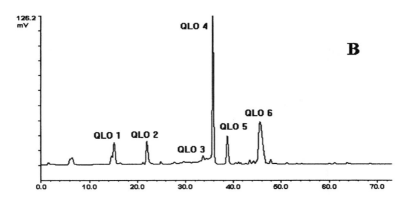

Figure 3 *RP-HPLC trace of quercetin degradation products under non-oxidative (A), and oxidative (B) conditions, after 240 and 80 min respectively, at room temperature (20 ± 1 °C). Chromatographic conditions as described previously. QLR 1 (QLO 1): 3, 4-Dihydroxybenzoic acid (protocatechuic acid); QLR 2 (QLO 2): 2, 4, 6-trihydroxybenzoic acid (phloroglucinol carboxylic acid); QLR 3 (QLO 3), QLR 4 (QLO 4), QLR 5 (QLR 5), and QLO 6: Unidebtified gedgradation products. Q: Quercetin*

much less great than during boiling. Thus, it becomes apparent that temperature is an essential factor in determining the nature of degradation products, under non-oxidative conditions.

Protocatechuic acid represents a fragment of the flavonol skeleton and derives from the B-ring, as shown in Figure 5. This reaction is greatly facilitated under vigorous conditions, *i.e.* reflux with 60% KOH[31] or 50% KOH in the presence of EtOH.[32] Nevertheless, it appears that generation of this compound occurs even when a much lower concentration is used (1.12% aqueous KOH), although there is a low yield especially during treatment at 20 ± 1 °C. Catechol was not detected during treatment at 97 °C, and therefore it was speculated that protocatechuic acid was not decarboxylated. However,

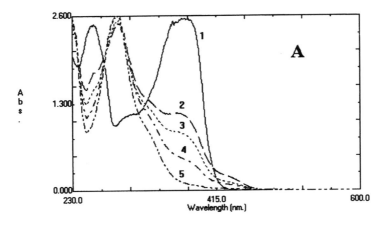

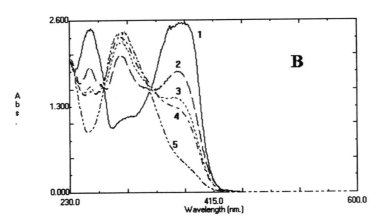

Figure 4 *Quercetin non-oxidative degradation at high (A) and low (B) temperature in aqueous medium (1.765) mM), at pH 23.0 ± 1, as shown by UV-Vix spectroscopy. Samples were diluted 1:10. 1: Quercetin in methanol (1.65 mM) at pH 3.0; 2, 3, 4, and 5: Samples taken after 30, 60, 120, and 240 min, respectively*

the concentration of this acid (based on peak area recorded at 260 nm) at 20 ± 1 °C after 240 min was almost 6-fold lower, although the difference in quercetin concentration was only *ca.* 5%. This implies that low temperatures may not facilitate protocatechuic acid formation. Phloroglucinol carboxylic acid was detected only during the low-temperature treatment at very low concentrations. Phloroglucinol was not detected at any time, and so it can be assumed that phloroglucinol carboxylic acid did not undergo decarboxylation either.

An early study[33] demonstrated that flavonols, under alkaline conditions and in the presence of a reducing agent, can be cleaved according to the mechanism

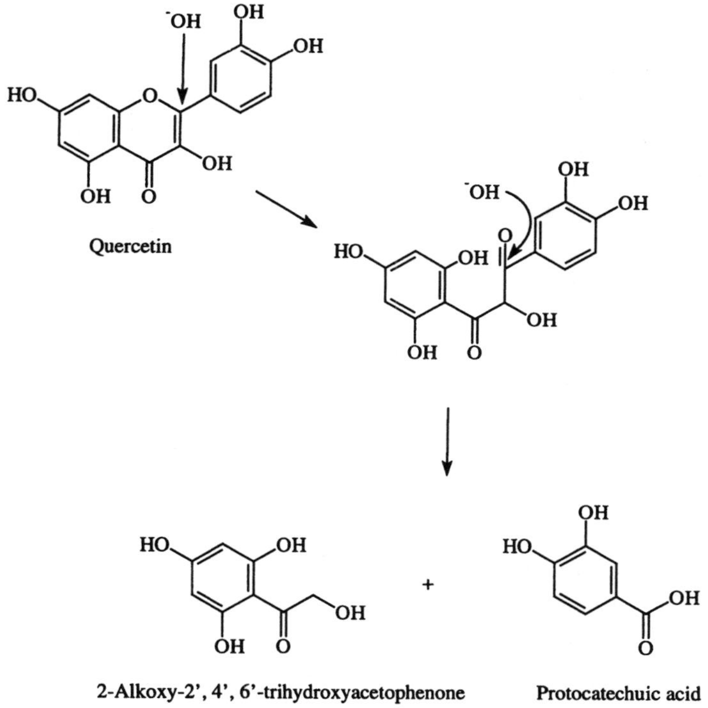

Figure 5 *Alkaline cleavage of quercetin (after Gottlieb, 1975)*

shown in Figure 6. This cleavage results in the formation of phloroglucinol (ex-A-ring), and in equal amounts of 3,4-dihydroxyphenylpropionic acid and its corresponding alcohol (ex-B-ring). However, this mechanism was inactive under the conditions employed since neither phloroglucinol nor any of the ex-B-ring fragments was detected. It can be said, therefore, that under non-oxidative conditions and in the absence of a reducing agent, flavonol thermal break-down in dilute alkaline media proceeds *via* the mechanism illustrated in Figure 5.

3.1.2 Oxidative conditions. Figure 7 shows the degradation kinetics of quercetin during boiling (97 °C) and at room temperature. As it can be seen, complete degradation was achieved after only 40 min, even during the low-temperature treatment. Thus it is apparent that the presence of oxygen greatly accelerated the rate of degradation, since complete decomposition of quercetin under non-oxidative conditions required more than 120 min. The fact that the low-temperature treatment had the same effect could be explained by the much greater solubility of oxygen in the alkaline solution at 20 ± 1 °C.

The HPLC profiles of degradation products at each temperature were

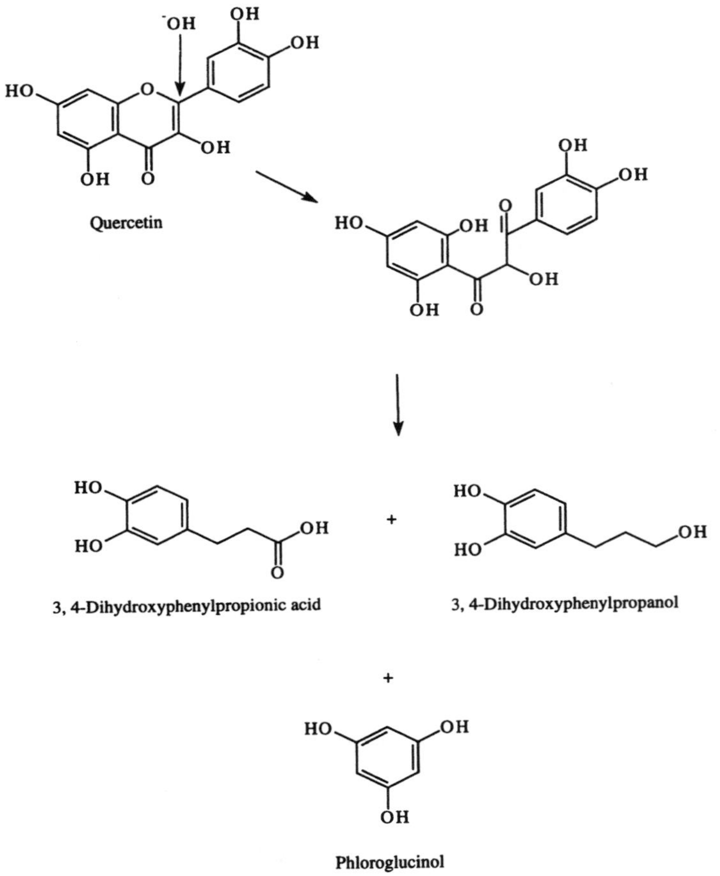

Figure 6 *Reductive cleavage of quercetin under alkaline conditions (after Hurst &
Harborne, 1967)*

essentially the same as those of the experiment obtained under non-oxidative
conditions. However, an additional product assigned as QLO 6, was formed
during treatment at $20 \pm 1\,^{\circ}\mathrm{C}$ (Figure 3B). This compound appeared as a
broad peak, was more polar than quercetin (in terms of a RP-HPLC
system), eluted after *ca.* 45 min, and its formation was accompanied by an
almost 4-fold increase in browning, as shown in Figure 8. This raises the
possibility that this compound may be polymeric in nature. It should be
mentioned at this point that polymeric products of unknown structure
deriving from quercetin autoxidation at alkaline pH have been reported,[34,35]
but no efforts have been made to isolate and identify these compounds.
Earlier studies,[36] have also provided evidence that autoxidation of flavonoids, such as catechin and related flavans in neutral aqueous solutions,
results in polymerisation.

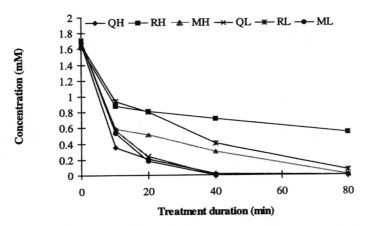

Figure 7 *Evolution of quercetin (Q), rutin (R), and quercetin/L-rhamnose/D-glucose (M) concentration during treatment under oxidative conditions. Letters H and L indicate treatment at high and low temperature, respectively*

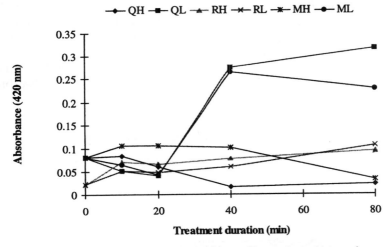

Figure 8 *Evolution of browning (A$_{420}$), during oxidative degradation of quercetin (Q), rutin (R), and quercetin/L-rhamnose/D-glucose (M). Letters H and L indicate treatment at high and low temperature, respectively*

The fact that this product was not formed at elevated temperature may be an indication regarding its thermostability. Alternatively, it can be postulated that its formation is dependent on product(s) that are formed only at low temperature. Moreover, it was observed that phloroglucinol carboxylic acid was formed in higher amounts than those found during the treatment under non-oxidative conditions. This finding is in accordance with previous reports,[37] which indicated that this substance is a minor product of oxidative cleavage of quercetin. Other researchers revealed that photo-oxygenation of quercetin can yield the depside protocatechuoyl phloroglucinol carboxylic acid,[38] which in turn can decompose giving protocatechuic and phloroglucinol carboxylic acid. This depside can also derive from quercetin through base-catalysed oxygenolysis[39] or microbial degradation.[40,41,42]

3.2 Examination of Rutin Solutions

3.2.1 Non-oxidative conditions. Rutin disappearance was linear between 30 and 240 min during boiling (97 °C) and at 20 ± 1 °C (Figure 1). After 30 min of boiling, only ≅ 41% of rutin was degraded, and the overall loss at the end of the treatment was ≅ 82%. The corresponding values at 20 ± 1 °C were 32% and 40%.

Browning was found to be considerably increased after 60 min of boiling, but declined towards the end of the treatment (Figure 9). This is possibly attributed to some products that appeared after 30 and 60 min, which did not survive until the end of the treatment. It is known (Theander, 1981; Eskin, 1991)[43,44] that when sugars are heated under alkaline (or acidic) conditions they give rise to reactions referred to as caramelisation. These reactions result in the formation of brown compounds of very complex chemical structure. Thus, it is probable that the products detected after 30 and 60 min (data not shown) were derived from decomposition of the sugar moiety, and were responsible for this transient increase in browning. However, the fact that these substances were not detected after 240 min of boiling suggests that they probably broke down to non UV-absorbing compounds.

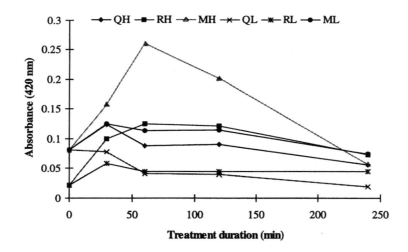

Figure 9 *Evolution of browning (A_{420}), during non-oxidative degradation of quercetin (Q), rutin (R), and quercetin/L-rhamnose/D-glucose (M). Letters H and L indicate treatment at high and low temperature, respectively*

At $20 \pm 1\,°C$ no particular change was observed and browning remained fairly constant. The latter correlates with the fact that during the same time period, rutin content decreased by only 8%. This is indicative of the resistance that the glycosidic bond confers to the flavonol molecule with respect to degradation. In addition, the HPLC analyses (Figure 2B) indicated that rutin degrades by a different manner to that of quercetin. The HPLC trace of rutin degradation products after 240 min of boiling was completely different than that of quercetin, the major product being a substance eluted after *ca.* 21 min. Its polarity, in terms of a RP-HPLC system, as well as its UV-Vis characteristics (max: 280, 308 nm; min: 251, 298 nm) indicated that it was not of flavonoid structure. Two other minor compounds assigned as RHR 3 and RHR 4, which were encountered among quercetin degradation products (Tables 3 and 5) were also detected.

Table 5 *Retention times of all major peaks detected during degradation of quercetin and rutin.*

Retention time (min)	Compound assignment				
15	QHR 1	RHR 1	QLR 1	QLO 1	RHO 1
21	RHR 2				
22	RHR 3	QLR 2	QLO 2	RLO 1	
23	RLO 2				
29	RLO 3				
33	QHR 5	RLO 4			
35	QHR 2	RHR 4	QLR 4	QLO 4	
37	QLR 3	QLO 3			
40	QHR 3	RHR 5	QLR 5	QLO 5	
43	RLO 5				
44	R				
46	QLO 6				
50	QHR 4				
66	Q				

QHR 1, RHR 1, QLR 1, QLO 1, RHO 1: Peaks identified as protocatechuic acid; QLR 2, QLO 2: Peaks identified as phloroglucinol carboxylic acid; R: Rutin; Q: Quercetin

Quercetin was not found at any time, and it might be said that hydrolysis of the glycosidic bond between rutinose and quercetin did not take place, since NH_4OH and H_2O_2 are required for such a reaction.[45] However, hydrolysis might have occurred to some extent, but quercetin could not accumulate to detectable amounts because it was shown that it is extremely labile under the conditions employed.

The degradation of rutin at low temperature proceeded by a different mechanism, as can be inferred from the HPLC trace after 240 min of treatment (Figure 10A). The breakdown resulted in more than fifteen products with very similar polarity. This caused a profound loss of resolution, and the UV-spectra of the compounds formed could not be obtained with reliability. However, comparing the two chromatograms, it can be supported that the degradation of rutin at low temperature did not yield the same products as the high-temperature treatment. Therefore, the formation of RHR 2, the major degradation product of rutin during boiling, seems to be exclusively dependent on heating. This also holds true for protocatechuic acid. Furthermore, the glycosidic bond between quercetin and rutinose appears as a critical factor in defining the mechanism of rutin degradation, since RHR 2 was not detected during quercetin degradation. The latter is confirmed by the fact that QHR 4 was not detected among rutin degradation products (Tables 3 and 5).

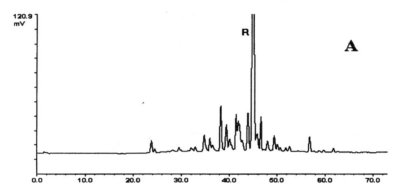

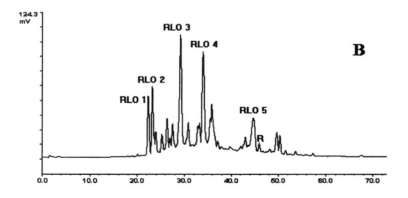

Figure 10 *RP-HPLC analysis of rutin degradation products under non-oxidative (A), and oxidative (B) conditions, after 240 and 80 min respectively, at room temperature (20 ± 1°C). Chromatographic conditions as described previously. RLO 1, 2, 3, 4, and 5: Unidentified degradation products. R: Rutin*

3.2.2 Oxidative conditions. The rate of rutin thermal degradation under oxidative conditions was, as in the case of quercetin, considerably increased compared to that achieved under non-oxidative conditions, and reached a level of 66.5% in 80 min (Figure 10B). The greater degradation observed at 20 ± 1°C (overall loss 99.5%), enhances the hypothesis that the greater solubility of oxygen at low temperature increases its concentration in the medium and accelerates degradation.

Browning increased 4- to 5-fold during treatment at both temperatures (Figure 8), probably due to caramelisation of the sugar moiety. Treatment at high temperature gave virtually the same HPLC profile of products obtained under non-oxidative conditions. RHO 3, along with protocatechuic acid and RHO 2, was also detected during the non-oxidative degradation of rutin, but in very minor amounts (Table 3). It is thus evident that, as for quercetin, the degradation mechanism of rutin at high temperature is the same under both oxidative and non-oxidative conditions, but in the presence of oxygen it is drastically accelerated. However, at $20 \pm 1\,°C$ rutin broke down to several peaks, five of them being the major degradation products (Figure 10B). The UV spectra of these compounds had λ maxima near 260 nm (Table 6) and did not correspond to products previously detected, thus suggesting formation of UV-absorbing compounds, possibly of aromatic nature, caused by decomposition of the sugar moiety.[44,46]

Table 6 *UV characteristics of rutin degradation products obtained under oxidative conditions at $20 \pm 1°C$*

Compound	UV data
ROL 1	max: 255 nm
	min: 241 nm
ROL 2	max: 260 nm
	min: 242 nm
ROL 3	max: 254, 304 nm
	min: 239, 281 nm
ROL 4	max: 259 nm
	min: 238 nm

One characteristic is that RHR 2, the main product formed during degradation of rutin at high temperature under both oxidative and non-oxidative conditions, was not detected. The possibility that other products previously found, either during quercetin or rutin degradation, might have formed at low concentrations cannot be ruled out, since their detection became particularly difficult due to lack of resolution. Furthermore, attempts to isolate and characterise any of the degradation products by means of semi-preparative RP-HPLC, preparative TLC and normal-phase column chromatography did not meet with success, because of their similar polarities. However, it is evident that compounds such as protocatechuic acid and phloroglucinol carboxylic acid were not formed, as these compounds have considerably shorter retention times.

It would be difficult to speculate about the degradation mechanism of rutin, due to lack of analytical data concerning the degradation products. However, from the UV characteristics presented in Table 6, it could be said that the cleavage of the flavonol skeleton is involved, since none of the products seems to be of flavonol nature. Nevertheless, it is noteworthy that rutin becomes very labile in the presence of oxygen even at low temperature, and it could be said that although the occupation of the 3-hydroxyl group was found to be essential for rutin to resist degradation under non-oxidative conditions, it could not sufficiently prevent rutin deterioration in the presence of oxygen and, possibly, in the presence of other oxidising agents. This is of particular importance because flavonol 3-glycosides are not ideal substrates for most plant polyphenol oxidases and peroxidases studied so far, but they may be cleaved during processing of plant foods and products, through mechanisms that do not involve enzymatic catalysis.

3.3 Examination of Quercetin/L-Rhamnose/D-Glucose Solutions

3.3.1 Non-oxidative conditions. The examination of rutin solutions provided sound evidence that the glycosidic bond between quercetin and rutinose does confer to rutin an increased resistance to degradation under both low and high temperatures, and influences significantly the degradation mechanism. However, the role of rutinose units, as such, in the degradation rate of rutin remained questionable, and for this reason solutions of quercetin containing equimolar amounts of L-rhamnose and D-glucose were also examined.

Apparently, the presence of sugars exerted a slight protective effect against quercetin break down, and the high-temperature treatment resulted in an overall loss of 95% (Figure 1), the corresponding value being 100% in the case of quercetin alone. This protective action was remarkably expressed at $20 \pm 1\,°C$, where only 53% of quercetin was degraded. It is noteworthy that the same treatment of quercetin alone yielded $\cong 95\%$ degradation. This is possibly attributed to the fact that higher temperatures result in rapid sugar degradation, while at lower temperatures sugar degradation is much slower, and therefore its protective action is extended. This may also explain the particularly high resistance of rutin against degradation at low temperature, compared to that of quercetin.

Another point that should be outlined is the approximately 4-fold increase in browning during the first 60 min of boiling (Figure 9). This obviously happened due to the decomposition of sugars, since such a phenomenon was not observed during treatment of solutions containing quercetin alone. It should be noted that browning exhibited similar behaviour during rutin degradation, and thus it is likely that browning reactions triggered by sugar decomposition are similar, irrespective of attachment of the sugars to the aglycone.

The presence of sugars did not influence the degradation mechanism, in general, and the degradation products both during high- and low-temperature treatments evolved, more or less, as in the case of solutions containing quercetin alone (data not shown). Nevertheless, it was observed that QHR 4 was formed at considerably lower concentration, whereas this compound was one of the major degradation products found after treatment of solutions containing quercetin alone. It could be postulated, therefore, that the sugars may also interfere with the degradation mechanism in an, as yet, unknown manner. Additionally, it is clear that there were no reactions between sugars and quercetin or any of its degradation products, since the substances detected were the same as those found during degradation of quercetin alone. This excludes the possibility that any of the products resulted from rutin degradation derived from reaction(s) between the sugar moiety and the aglycone.

The above findings indicate that the two sugar units, although not attached to the aglycone, were able to alter the kinetics of quercetin degradation and the formation of degradation products, under non-oxidative conditions. Furthermore, the data obtained provided evidence that the high resistance of rutin to degradation is not purely attributed to the glycosidic bond, but also to the sugar moiety as such.

3.3.2 Oxidative conditions. The presence of sugars obviously retarded quercetin degradation at high temperature, although the overall loss at the end of the treatment was 99.5%. Once again, however, treatment at $20 \pm 1\,°C$ proved to be more efficient (Figure 7). This effect was observed for both quercetin and rutin and seems to be of particular importance, since a rapid deterioration could be achieved in a short time even at a low temperature.

The HPLC traces of the products formed during quercetin degradation in the presence of L-rhamnose and D-glucose at both temperatures, were very similar to those obtained during quercetin degradation. This excludes the possibility of reaction(s) between the sugars and quercetin or any of its degradation products.

Therefore, it can be stated that the free sugar units, although they played a substantial role in determining both the rate of quercetin degradation and formation of degradation products in the absence of oxygen, had an insignificant effect under oxidative conditions, especially at low temperature. Additionally, sugars do not seem to contribute towards the formation of compounds detected during rutin degradation, thus highlighting the importance of the glycosidic bond between quercetin and rutinose.

4 Discussion

Flavonol thermal degradation is important for two reasons. Firstly, the behaviour of flavonols in aqueous media at high temperatures is a crucial factor in assessing their potency for commercial use as natural antioxidants in foods. This is particularly interesting because flavonols are powerful antioxidants but their thermal stability, as well as their interactions with other food components, are poorly studied. In addition, although it is known that polyphenols are greatly involved in the development of food browning, the role of flavonols and flavonol glycosides in similar reactions has not been investigated in details. Secondly, the fate of flavonols in plant foods undergoing industrial or domestic processing is a significant issue in terms of nutrition, since this class of flavonoids is now considered as being essential constituents of plant foods because of their multiple biological activities. Furthermore, the nature of flavonol degradation and reaction products, as well as their biological activities, remain obscure.

Although it may be said that the conditions under which the experiments were carried out are not frequently encountered during processing, it is likely that the results of this study will provide data that may also be obtained under real conditions. It was reasoned that a detailed study of a flavonol and a glycoside thereof, under alkaline conditions, would serve as a means of guiding further studies under conditions more typical for foods. To the extent that the reactions and mechanisms proposed should be similar, the results may be directly applicable.

From the examinations carried out, it was highlighted that quercetin degradation in aqueous media under alkaline conditions is greatly facilitated in the presence of oxygen, especially at low temperatures where oxygen solubility in the medium is high. At higher temperatures, such as those that are normally attained during boiling, degradation yields the same products under both oxidative and non-oxidative conditions. At room temperature and under oxidative conditions, a dramatic increase in browning is produced, possibly due to the formation of a compound which is not encountered among the degradation products of quercetin at high temperature.

As far as it concerns the degradation of rutin (quercetin 3-O-rhamnosylglucoside), the presence of oxygen was found to be once again essential in achieving particularly rapid degradation rates, both at low and high temperature. Under non-oxidative conditions, the glycosidic bond between quercetin and rutinose gives rutin an exceptional resistance to degradation, especially at low temperature. However, similar resistance at low temperature can be achieved even if the two sugar units of rutinose, L-rhamnose and D-glucose are not glycosidically bound to the aglycone. This effect, nevertheless, is almost completely abolished under oxidative conditions.

The glycosidic bond is able to alter substantially the degradation mechanism and kinetics observed for quercetin, but this is also dependent to some extent on the presence of sugar units as such. In addition, dehydration and/or fragmentation of the sugar moiety appear as possible causal factors for the formation of the products detected during rutin degradation under both non-oxidative and oxidative conditions. Quercetin in the presence of L-rhamnose and D-glucose did not give rise to different compounds, and thus it is obvious that the glycosidic bond between quercetin and rutinose is essential in determining the formation of rutin degradation products. However, the finding that thermal degradation of both quercetin and rutin yields protocatechuic acid, RHR 3 (QHR 2) and RHR 4 (QHR 3) may be an indication that cleavage of the flavonol skeleton has common features in both cases.

As it can be easily understood, flavonols and especially flavonol glycosides could be considered fairly stable molecules in an environment that lacks oxygen, this stability of the aglycones being even greater in the presence of sugars. It can be assumed, therefore, that in food systems there may be other compounds that could exert protective effects on flavonols, with respect to degradation. On the other hand, it is evident that oxidative conditions convert flavonols into compounds that are prone to deterioration, even at ambient temperature. In this case the role of sugars, either free or conjugated with the aglycone, is rather insignificant.

It is anticipated that the identification of the remaining degradation products, which is currently in progress, will further elucidate the degradation mechanisms of both quercetin and rutin and will provide evidence concerning the changes to flavonols and, probably, to other flavonoids during thermal processing of plant foods and products.

Acknowledgements

The authors would like to thank Keith Price and John Eagles (Institute of Food Research, Norwich), for their assistance in mass spectrometric studies.

References

1   J. Kühnau, *World Rev. Nutr. Diet*, 1976, **4**, 117.
2   K. Herrmann, *J. Food Tech.*, 1976, **11**, 433.
3   A. Bilyk and G.M. Sapers, *J. Agric. Food Chem.*, 1985, **33**, 226.
4   M.G.L. Hertog, P.C.H. Hollman and M. B. Katan, *J. Agric. Food Chem.*, 1992, **40**, 2379.
5   A. Crozier, M.E.J. Lean, M.S. McDonald and C. Black, *J. Agric. Food Chem.*, 1997, **45**, 590.
6   A. Finger, U.H. Engelhardt and V. Wray, *J. Sci. Food Agric.*, 1991, **55**, 313.
7   M.G.L. Hertog, P.C.H. Hollman and B. van de Putte, *J. Agric. Food Chem.*, 1993, **41**, 1242.
8   M. Toyoda, K. Tanaka, K. Hoshino, H. Akiyama, A. Tanimura and Y. Saito, *J. Agric. Food Chem.*, 1997, **45**, 2561.
9   P.C.H. Hollman, J.H.M. de Vries, S.D. van Leewen, M.J.B. Mengelers and M.B. Katan, *Am. J. Clin. Nutr.*, 1995, **62**, 1276.
10 C. Manach, C. Morand, O. Texier, M.-L. Favier, G. Agullo, C. Demigné, F. Régérat and C. J. Rémésy, *J. Nutr.*, 1995, **125**, 7, 1911.
11 E. Corvazier and J. Maclouf, *Biochim. Biophys. Acta*, 1985, **835**, 315.
12 R.L. Sousa and M.A. Marletta, *Arch. Biochem. Biophys.*, 1985, **240**, 1, 345.
13 R.J. Gryglewski, R. Korbut, J. Robak and J. Swies, *Biochem. Pharm.*, 1987, **36**, 3, 317.
14 C.R. Pase-Asciak, S. Hahn, E.P. Diamandis, G. Soleas and D.M. Goldberg, *Clin. Chim. Acta*, 1995, **235**, 207.
15 A. Lale and J.M. Herbert, *J. Nat. Prod.*, 1996, **59**, 273.
16 R.L. Divi and D.R. Doerge, *Chem. Res. Toxic.*, 1996, **9**, 16.
17 P. Cos, L. Ying, M. Calomme, L.P. Hu, K. Cimanga, B. van Poel, L. Pieters, A.J. Vlietinck and D. Vanden Berghe, *J. Nat. Prod.*, 1998, **61**, 71.
18 B. Havsteen, *Biochem. Pharm.*, 1983, **32**, 7, 1141.
19 K.R. Price, J.R. Bacon and M.J.C. Rhodes, *J. Agric. Food Chem.*, 1997, **45**, 938.
20 S. Hirota, T. Shimoda and U. Takahama, *J. Agric. Food Chem.*, 1998, **46**, 3497.
21 K.R. Price, F. Casuscelli, I.J. Colquhoun and M.J.C. Rhodes, *J. Sci. Food Agric.*, 1998, **77**, 468.
22 S. Jood, B.M. Chauhan and A.C. Kapoor, *J. Sci. Food Agric.*, 1987, **39**, 145.
23 P.U. Rao and Y.G. Deosthale, *J. Sci. Food Agric.*, 1982, **33**, 1013.

24   C.-F. Chau, C.-K. Cheung and P. Wong, *J. Sci. Food Agric.*, 1997, **75**, 447.

25   L. Coward, M. Smith, M. Kirk, andS. Barnes, *Am. J. Clin. Nutr.*, 1998, **68** (suppl.), 1486S.

26   S.M. Mahungu, S. Diaz-Mercado, J. Li, M. Schwenk, K. Singletary and J. Faller, *J. Agric. Food Chem.*, 1999, **47**, 279.

27   J. Raynal and M. Moutounet, *J. Agric. Food Chem.*, 1989, **37**, 1051.

28   J. Raynal, M. Moutounet and J.-M. Souquet, *J. Agric. Food Chem.*, 1989, **37**, 1046.

29   J. A. Larrauri, P. Rupérez and F. Saura-Calixto, *J. Agric. Food Chem.*, 1997, **45**, 1390.

30   H.P.S. Makkar and B. Singh, *J. Sci. Food Agric.*, 1991, **54**, 323.

31   R.M. Horowitz and B. Gentili, *J. Org. Chem.*, 1961, **26**, 2899.

32   G. Berti, O. Livi, D. Segnini and I. Cavero, *Tetrahedron*, 1967, **23**, 2295.

33   H.M. Hurst and J.B Harborne, *Phytochemistry.*, 1967, **6**, 1111.

34   C.G. Nordström and C. Majani, *Suom. Kemist.*, 1965, **38**, 11, 239.

35   Q.H. Nguyen, M. Metche and E. Urion, *Bull. Soc. Chim. France*, 1966, **7**, 2232.

36   D.E. Hathway and J.W. Seakins, *J. Chem. Soc.*, 1957, 1562.

37   O.R. Gottlieb, Flavonols, in *The Flavonoids*, ed. J.B. Harborne, T.J. Mabry and H. Mabry, Chapman & Hall, London, 1975.

38   T. Matsuura, H. Matsushima and R. Nakashima, *Tetrahedron*, 1970, **26**, 435.

39   A. Nishinaga, T. Tojo, H. Tomita and T. Matsuura, *J. Chem. Soc., Perkin Trans. 1*, 1979, 2511.

40   S. Hattori and I. Noguchi, *Nature*, 1959, **184**, 1145.

41   D.W.S. Westlake, J.M. Roxburgh and G. Talbot, *Nature*, 1961, **189**, 510.

42   H.G. Krishnamurty and F.J. Simpson, *J. Biol. Chem.*, 1970, **245**, 6, 1467.

43   O. Theander, *Progr. Food Nutr. Sci.*, 1981, **5**, 471.

44   M.N.A. Eskin, *Biochemistry of Foods*, Second edition, Academic Press, 1991.

45   K.R. Markham, Flavones, flavonols and their glycosides, in *Methods in Plant Biochemistry*, Vol. 1: Plant Phenolics, ed. J.B Harborne, Academic Press, 1989.

46   I. Forsskåhl, T. Popoff and O. Theander, *Carbohyd. Res.*, 1976, **48**, 13.

Subject Index